Marine Electronics

The Straightshooter's Guide to

Marine Electronics

2nd edition

Gordon West and Freeman Pittman

with illustrations by Jim Sollers

Distributed by:
Airlife Publishing Ltd.
101 Longden Road, Shrewsbury SY3 9EB, England

International Marine Publishing Company
Camden, Maine 04843

Published by International Marine Publishing Co., a division of Highmark Publishing Ltd., 21 Elm Street, Camden, Maine 04843

Typeset by Camden Type 'n Graphics, Camden, Maine.
Printed and bound by Alpine Press, Stoughton, Massachusetts.
Text and cover design by Ken Gross.

10   9   8   7   6   5   4   3   2   1

Library of Congress Cataloging-in-Publication Data

West, Gordon.
The straightshooter's guide to marine electronics.

1. Boats and boating—Electronic equipment.
I. Pittman, Freeman.   II. Title.
VM325.W46   1987   623.8'504   87-3563
ISBN 0-87742-241-9

Contents

Preface

If we have been successful, what follows is an objective and to-the-point survey of the marine electronic instruments and systems commonly used by sail and powerboat owners and by sport and commercial fishermen. Our goal is to give you the information you need to judge what is out there and what is compatible with your boat and your purposes. Along the way, we will explain the operation, capabilities, and limitations of each instrument.

We'll discuss the ins and outs of installing marine electronics in your boat, and we'll examine a related consideration—whether to buy from a local marine electronics dealer (who will probably install the equipment as part of a sales package) or a mail-order discount house (who will not). We'll also include some procedures for maintaining the gear and for troubleshooting malfunctions. Too many mariners are frightened into inaction by the complex circuitry contained in most instrument housings. Better than 50 percent of all marine electronic malfunctions could be fixed by anyone with a little patience and common sense, and by maintaining your own equipment and doing the simpler repairs yourself (as well as the installation, tuning, and calibrations, when possible), you save money, avoid the inconvenience of equipment "down time," and become a more skilled and knowledgeable operator.

This guide does not attempt model evaluations. To do so would have required independent testing facilities, and in any case, the marine electronics marketplace is evolving so rapidly that at least some such comparisons would have been out of date by the time the book was published. Instead, we have attempted to provide all the information you will need to compare and evaluate models yourself in an informed fashion. This knowledge should be as current in 1988 as it is in 1987.

By using the charts in Appendix C, which contain complete manufacturer's specifications and prices for depthsounders, VHF radiotelephones, cellular phones, and autopilots, you can start to identify the electronics that will best suit your needs, be they related to fishing, racing, daysailing, or cruising. Since manufacturers are always adding new models to their lines, you may wish to narrow your search through the charts provided, then write manufacturers directly for their latest product listings. Names and addresses of manufacturers, importers, and distributors are given in the back of the book for easy reference. Prices in our charts are "suggested," and you can generally do better.

Perhaps the most important consumer information in this book concerns the question of where to buy a marine electronic item or package. The most frequent choice is either a local marine electronics specialty dealer or a mail-order discount establishment. The former are too numerous to list; you can probably find a good one in your locale. Chapter One gives you some criteria to go by, but don't forget the importance of word of mouth. The mail-order discount houses are listed at the back of the book, along with the manufacturers of the equipment we discuss. You can write the latter directly for information on gear that interests you.

Never before have we had so much good marine electronic equipment to choose from. This equipment is much smarter, more compact, and in many instances *less* expensive now than it was just a few years ago. Not surprisingly, the process of shopping for gear is correspondingly more complicated. We hope this guide helps.

1

The Marine Electronics Marketplace

by Gordon West

Probably the most frequently asked question in the marine electronics business is, "Which unit is best?" It's a good question, but there is no one right answer. The question should perhaps be rephrased as, "Which marine electronics are best for my particular application?" To answer this properly, the marine electronics seller would need to ask you many questions about your boat, your intended cruising or fishing area, your experience in operating electronic equipment, and the level of sophistication you desire. If any seller tells you which unit is best without interrogating you first, look for another place to shop.

Deciding Where to Buy

There are four major avenues of distribution for marine electronic equipment:

1. The traditional marine electronics dealer
2. Marine electronics catalog houses
3. Marine hardware stores that also sell electronics
4. Boat dealers who include built-in electronics with new or used boat sales

There is no love lost between the specialty dealers and the catalog houses. The sales and servicing of marine electronics have traditionally gone exclusively to the former, but the catalog houses have recently made major inroads on the dealers' sales.

Most specialty dealers have elaborate showrooms with examples of every type of marine electronic gear hooked up and operating. This allows you to hear, manipulate, and get the feel of any type of equipment you may be thinking of purchasing, and there is no substitute for this. Most dealer showrooms also stock an ample supply of literature for each piece of equipment that is on display, as well as some that are not.

The specialty dealers hire the cream of the crop when it comes to qualified sales personnel. Many of the men and women who make up their sales forces are qualified not only to recommend gear but to give you a fairly technical overview on any piece of equipment in their showroom. In addition to these sales personnel, a dealer usually employs three to eight FCC-licensed technicians who possess marine electronics technician certificates; these are the top experts in their field, and many of them have had extensive additional factory training on specific products. (Equipment manufacturers spend thousands of dollars training technicians on the installation and operating techniques and the servicing of particular pieces of equipment.) If you have a question that the dealer's sales force can't answer, they can often bring out the technician "from the back" who will answer your query in detail. The dealer can also provide you with a custom installation aboard your boat, or assist you with a do-it-yourself installation.

The marine electronics specialty dealers also offer factory-authorized warranty repair. Thus, if your new piece of gear should fail, it may be fixed for free by the dealer. Sometimes the equipment must be taken back to the shop for service, but in other cases the fix can be accomplished aboard the boat. Troubleshooting of equipment aboard your boat is another unique feature of a specialty dealer. They have the portable equipment to do it. Labor rates are high if not covered by warranty, but are usually worth every penny. (More about warranties shortly.)

Specialty dealers have united with manufacturers to form a trade organization called the National Marine Electronics Association (NMEA). The NMEA has a code of ethics that all member dealers must meet; if a dealer sports the NMEA logo, you can be sure that he is indeed a specialty dealer who has met the membership requirements.

Marine dealers normally sell their equipment at the manufacturers' suggested retail prices. They are not known for heavy discounting, for they must maintain a minimum markup in order to cover the showroom, the trained sales personnel, and the expensive service equipment. It is for this reason that catalog sales of marine electronic equipment have been growing so strongly in recent years.

Marine catalog houses that sport exciting color brochures and even more exciting discount prices are indeed doing well. These relative newcomers to the world of marine electronics got their start when gear that could be user-installed became more common, and when the manufacturing of marine electronic gear began to overwhelm the available avenues of distribution. Twenty years ago most marine electronic equipment was American-made, and production was just right to support the specialty dealers. There was no such thing as too much gear to sell, and discounting was taboo. Now United States manufacturers of marine electronics are importing a huge portion of their gear from the Far East. In doing so, however, the domestic manufacturers have lost control of production volume. The Oriental manufacturers don't and won't slow down on production, so alternate sources of fast distribution are necessary to keep the pipeline clear. The name of the game for the marine electronics catalog-discount house is quick sales, which they achieve with a low price. Some very aggressive promotional work is carried on by the marine electronics discount catalogs and showrooms. Their creative marketing skills leave the specialty dealers in the dust. Manufacturers must surely be impressed with their sales figures, as well as their promotional ideas.

If you plan to install your equipment yourself, and you know what you want, discount houses and catalogs might be a good way to go. But don't expect many catalog houses to give you an hour or two of their time going over a particular piece of marine gear. Their toll-free numbers are for placing orders, and their personnel, though they may indeed be knowledgeable about marine electronics, are not on the phone to plan your entire marine electronics system. Be advised, too, that discount and catalog houses frequently don't tune or repair the equipment they sell, and most don't offer any installation service at all. Advice, yes; service, no. Several marine electronics manufacturers will not include the necessary FCC certification with their radiotelephones or radar, and a factory-authorized and licensed dealer must either install the equipment for you or inspect your installation prior to "signing off" on the FCC radar or radiotelephone certificate. If you buy the equipment from a discount house, this will be an additional out-of-pocket expense. If your equipment breaks, most catalog houses will tell you to ship it back to the manufacturer for repair. This "we sell it, they fix it" approach may not be popular with manufacturers, but when a lot of units are sold through these houses, they don't have much choice.

Purchasing a piece of marine electronics through a discount house, having it fail, and then taking it to the marine electronics specialty dealer down the street for repair may cause hard feelings, especially if the specialty dealer took the time originally to tell you which piece of equipment to buy. Some dealers might throw you out the door. Others, if they have a repair shop, may fix your unit, but most will take their sweet time if you bought it elsewhere.

A few marine electronics discount organizations do possess service as well as installation facilities. These are rare, however, and you should carefully check out such an organization before buying equipment from them. You may wish to use the checklist at the end of this chapter for that purpose. Even if a discount house staffs two or three technicians, these are probably unable to keep up with the repair work generated by the discounter's large volume of sales. As a general rule, when you buy equipment from a catalog house, you will encounter less delay and frustration if you send it directly back to the manufacturer for repair.

Buying electronics in a marine hardware store is downright questionable. What you will find will be a gaggle of equipment sealed under the glass in the showroom counter. Rarely is it hooked up, and seldom will the hardware store personnel know anything about it. If the equipment should fail, they won't be able to service it. They will usually send you down the street, where you should have bought it in the first place. Marine electronic gear is getting as sophisticated as a personal computer. Let the hardware stores sell their cleats, lines, and boathooks, and don't buy marine electronics through any store that can't tell you how to install, run, or fix them. There may be some exceptions, but most hardware stores have no expertise in this area.

Buying marine electronic equipment with a boat and having it installed by the boat dealer is another proposition. Most boat dealers will subcontract the installation to local marine electronics specialists, so you can get a good deal in a packaged boat sale. Just be sure to find out who's going to do the installation of the equipment, who will service it, and who in the world is going to show you how to run it. Most boat dealers are specialists in boats, not electronics.

If you are buying a boat that already has marine electronics aboard, make a list of the gear and take it to a local marine electronics specialist to find out how up to date it is. The better marine electronics specialists will give you a fair answer, but you may want a second opinion. New boats with built-in marine electronics may be a very good deal if the electronics are exactly what you are looking for. The more serious mariner or marine electronics enthusiast, however, will specify certain pieces of equipment. If such a request can be honored, make sure a qualified technician will do the installation.

The retail margin from the manufacturer to the seller of marine electronics averages about 40 percent. On large equipment such as radars, water makers, and sonars, it is usually less, and on smaller pieces of equipment such as VHF transceivers, depthsounders, and RDFs, the margin can be as much as 50 percent. Marine antennas may go as high as 60 percent. Keep this in mind when you are offered a percentage discount off the manufacturer's suggested retail price.

The marine electronics specialty dealers like to operate at a 30 to 40 percent margin, while the discounters can operate as low as 10 to 15 percent above cost if they move large quantities. Boat dealers that offer marine electronics as a package may sometimes break even—the marine electronics will help them sell the boat, and that's their main interest. This gear will not be too heavily discounted, however, since the boat dealers buy from another distributor rather than directly from the manufacturer. The hardware stores, too, usually buy from distributors, the distributor making about 10 percent and the hardware store hoping to sell the gear at a margin of 30 percent.

Summing up, if you want a turnkey operation and the ability to put your finger on one person to select the appropriate equipment, install it, and show you how to use it, the marine electronics specialist is a good way to go. If you want the best buy for the least money and you plan to install and tune everything yourself in your spare time, you might want to buy your gear through a discount house or catalog.

If it's your wife's birthday and a Sunday to boot, and she absolutely has to have that marine VHF handheld radio, and you spot it in an open marine hardware store, then go ahead—pay the retail price, get your set, and hope it doesn't break.

If the boat price is no longer negotiable but you can get the dealer to throw in a bucketful of marine electronics, have at it.

Equipment installation is an important factor. Unless you're a confirmed do-it-yourselfer or an electronics hobbyist, for example, you're probably going to want to have your radar system, single sideband station, or automatic direction finding station professionally installed aboard your boat. If you buy the equipment from a mail-order house and then ask a local marine electronics dealer to do the installation, you'll be charged on a time-and-materials basis. Using SSB radio as an example, a typical installation takes two days or more, and at $40 per hour plus materials you would be facing an additional out-of-pocket expense of at least $700.

If, after the unit was installed and certified, you were not satisfied with its operation, the additional grounding or other recommended remedy would cost you more money still. If, on the other hand, you buy the SSB radio (or radar, or ADF station, or loran, or . . .) from the local dealer, you can have the price of the installation included as part of a sales-and-service package. Have this agreement drawn up beforehand, in writing, and insist that it permit you to call back the technician *without additional charge* if you are not happy with the initial installation. An installation included as part of a sales agreement should cost substantially less than one contracted on a separate, time-and-materials basis.

Probably the best approach is to buy larger, hard-to-install, hard-to-remove gear from the marine electronics specialist and smaller, removable equipment from a discount house. If the big stuff breaks you can bring the specialist aboard to fix it. If the little stuff goes sour, you can unplug it and either take it into the discounter's local repair depot or ship it back to the factory. The discussions of individual instruments in this book will give you a feel for which you may safely buy from a discounter.

The most important thing to remember, no matter where you buy your equipment, is not to trust anyone who says, "This equipment is best for you," without first finding out how you plan to use it. Take time to look at your options when buying marine electronics, and always play fair—don't pick the brain of the marine electronics specialist only to turn around and buy your gear through a discount house. Stay honest with the people you are dealing with, and you may find that they will meet you halfway when you tell them what you want to put aboard and what your budget is.

One further warning: No matter who you buy from, you are definitely taking a chance when the new equipment you buy is delivered in a carton that has been previously opened without any documentation having been added. Some sellers check out and certify equipment before selling it, and if this is the case, that's fine. But demand their certification sheets on the equipment in the preopened box. Otherwise, you may be getting a unit that has already been sold once. It is not unusual for a buyer to return a unit to a dealer complaining of one defect or another. If the dealer can't find the defect

on the test bench, he may decide the buyer was mistaken, repack the unit, and sell it to the next customer, who might be you.

Warranties

Warranty periods for free repair and possible retuning of the equipment you buy may be as short as 90 days or as long as two years. Study the warranty before you buy, because there are plenty of loopholes.

The manufacturer's warranty is expressed clearly enough in the paperwork that comes with each piece of equipment. If you take the unit off your boat and send it back to the manufacturer (who is in reality the American importer if the unit is manufactured in the Orient), they will fix it free of charge provided the warranty has not expired. You pay the shipping charges, however. Most factories will get the unit back to you within three weeks. But if it's not convenient or even feasible to return a piece of gear to the manufacturer for repair, you will be dependent on a factory-authorized, licensed repair facility, and here you may encounter delays or frustration.

As discussed, not all sellers of marine electronic equipment are licensed repair facilities. If you buy from a discounter, for example, you may have to go somewhere else to have the unit repaired or tuned. If that somewhere else is a specialty dealer, he may or may not do the repair, and he will almost certainly put your gear at the end of the line behind any equipment bought directly from him. He will probably refuse to honor the factory-authorized warranty if you do not have proper sales slip information, so do send in the warranty cards and do keep accurate purchase records.

Marine electronics specialty dealers and a few discount catalog sellers will service everything they sell, and are factory-authorized to do so. They will give you a written statement on the terms of their free service, and they may have extended-service options as well. These written statements should be examined carefully. While manufacturers will often pay dealers to service your equipment when it is under warranty (many demand that the more complex units, such as SatNav receivers, be returned to them for repair), they may not pay a technician's travel expenses to your boat. Will a dealer pay for that service if you buy the unit from him? Find out beforehand; a permanently installed radar unit would not be convenient to dismantle and return for service.

Remember, too, that even if a discount catalog seller is factory-authorized to service and repair the equipment, you will usually pay the shipping charges, and you would probably get your equipment back sooner were you to ship it directly to the manufacturer.

When you buy equipment, ask what will happen should your gear fail in a distant port. Can the seller or manufacturer supply you with a list of authorized service facilities that will honor your warranty? For reasons already stated, this is particularly important when you buy mail-order equipment, but it applies when you buy from a specialty dealer, too. Although specialized marine electronics dealers are united through the NMEA, a dealer 1,000 miles away from where you bought the unit will not necessarily honor the warranty and fix your set for free.

Get your warranty claims in writing, and more important, find out exactly what the local marine electronics dealer will do to back up the manufacturer's warranty. Some enterprising dealers will exchange any item that fails within the first 10 days of operation. That's a nice feature to look for.

Comparing Models

Comparison charts of manufacturers' model specifications are included in Appendix C to help you in your search or even your daydreaming. The marine electronics marketplace changes so rapidly, however, that these charts will begin to slip out of date from the month they are published, although their usefulness should be largely unaffected for two years or more from publication.

Most of the changes in a given type of marine electronic gear are small and incremental, however, and the fundamental criteria by which the gear should properly be judged change more slowly than the gear itself. It is on that premise that the consumer advice in this book is based. For each marine electronic instrument or system, the criteria of selection are set forth. Follow these criteria carefully,

supplementing the process with input from dealers and fellow mariners, and the gear you buy will be appropriate for your purposes.

The National Marine Electronics Association gives annual awards to the "best piece of marine electronics" in each category, these awards being decided by the dealers. But although the quality and capabilities of the equipment are weighed in the selection process, the dealers are biased toward units that give them a good profit margin, and the awards are thus, in part, a popularity contest. Also, marine electronics dealers have been slow to accept new and innovative designs, preferring to wait as long as several years while the designs prove themselves. Ten years ago, when the first keyboard-entry VHF marine radio was introduced, it represented a breakthrough, but it won no industry awards or accolades partly because the marine electronics specialists simply didn't trust a radio without knobs, and partly because they wanted a more favorable profit margin on the product. It is true that new designs sometimes need a year or two to stabilize and have their bugs worked out (and you can use this as a criterion for selection if you don't mind buying last year's technology), but there are nevertheless new instruments each year that deserve awards but garner none. The NMEA awards signify marketability first and foremost, and true excellence only incidentally.

The quality of the instruction manual is a good clue to the quality of the marine electronics unit you may be planning to buy. Better units will have a more complete instruction book, and some units will have an additional service manual that may be purchased separately. If you can't quite decide which piece of electronics to choose, ask the dealer for a peek at the instruction book.

A technical service manual is a boon if you plan to have the equipment serviced by different dealers along the coast. This eliminates the age-old problem of dealers telling you they can't fix your equipment because they don't have the appropriate service manual. The service manual might also give you some basic troubleshooting guides to follow should your equipment go on the blink. Again, the more complete the service manual, generally the better the equipment.

Most important, however, the best electronics are those which fit with greatest precision your installation and operating needs. Again, if a seller recommends a particular model before asking you any questions about how you plan to install and use it, he is only selling equipment, and not a service. Don't decide on the name brand until you have seen them all and reviewed the capabilities of each.

Buyer's Checklist

Is the marine electronics seller you are considering characterized by the following?

- Factory-authorized service approval
- Sells marine electronics as its major business
- Buys directly from the manufacturer
- Offers trained salesmen to spend up to one hour with you in planning your system
- Has demonstration equipment on the floor
- Offers catalog sheets for you to study at home
- Has a licensed technician available for consultation
- Offers installation and service aboard your boat
- Possesses the necessary technical equipment and personnel to calibrate, tune, test, and FCC-certify all equipment
- Offers a selection of equipment
- Stocks the popular accessories for each unit
- Boasts factory achievement certificates of technical merit for technicians

Here are some other questions to ask:

- Is the equipment sold to you in factory-sealed containers, or is it "demo gear" taken home by someone else?
- Can the seller modify the equipment to meet particular installation needs?
- Will the seller replace dead-out-of-box equipment?
- Will the seller loan you a piece of equipment while your new piece is getting fixed?
- Can the seller provide a custom installation, such as cabinetry, for your equipment?
- Is the seller a member of the NMEA?
- Is the showroom or catalog tidy, complete, and professional?
- Has the seller been in business for more than a few years?
- Can your purchased equipment be fixed in any other cruising area in the United States or the world?
- Is the pricing competitive?

If you can answer "yes" to most of these questions, you have found a good, safe place to buy. For small pieces of equipment, if you know what you want, you can accept "no" answers to more of these questions as you gravitate toward reputable discount houses in search of a better price. If you *do* buy from a dealer, demand the strong warranty backup for which you're paying, and demand that the loran-C receiver, SSB radio, SatNav receiver, radar set, autopilot, or automatic direction finder be properly tuned, calibrated, tested, installed (if dealer installation is part of the package), and, if necessary, FCC-certified.

2

Depthsounders

by Gordon West

There have been huge advances in depthsounder technology within the last five years. Foreign and American depthsounder manufacturers are battling it out to see who can come up with the most features at the best price, and we consumers reap the benefits. The latest generation of depthsounders will work at higher boat speeds, suffer less ignition interference, ignore swiftly passing air bubbles, and stay locked onto the bottom for up to five seconds when your boat pops completely out of the water at 50 knots. Yet prices are lower than they have ever been. Two distinct types of depthsounder have emerged, one for navigation and one for locating fish. This chapter covers both.

How They Work

The principle of operation behind echo sounding hasn't changed over the last 60 years. A depthsounder beams pulsed sound waves down toward the bottom, then picks up the weak echoes and converts this information into bottom and fish readout information, which is displayed on the indicator. While variations in water temperature, salinity, and density affect the speed of sound some-what, acoustical pulses normally travel at about 4,945 feet per second in water, compared to 1,117 feet in air. If it takes two seconds for a pulse to leave the underwater sender/receiver unit, bounce off the bottom, and return to the unit, the depth of water must be 4,945 feet.

Depthsounders consist of three components:

1. The *transducer*. This is the underwater device that translates pulses of electrical energy into acoustical vibrations, directs these toward the bottom, receives the echoes, and converts them back into electrical impulses. The transducer case may be all plastic or bronze with a plastic face, and it is mounted with its face in the water and pointed downward.

2. The *amplifier*. Located inside the depth indicator unit, this component actually comprises three subcomponents: an oscillator to produce electrical impulses of a uniform frequency; a transmitter amplifier to step up the impulses before sending them to the transducer; and a receiver amplifier to pick up the faint return echoes and convert them into strong, usable signals that are fed into the display or indicator circuitry. Bigger and

more expensive transmitter amplifiers give stronger signals to reach extremely deep bottoms.

3. The *indicator*. Take your pick! Chart recorders, flashing neon bulbs, liquid crystal displays, video presentations, and other readouts are available for the indicator unit. This is the device that translates the transducer input into a readout form—graph, digital, or analog—that you can understand and interpret. When you select one of the several types of sounders, what you are really choosing among is the various types of indicator units.

The transducer is a critical part of your depthsounder assembly. Many different types are available with any one type of sounder, and the one you choose must fit your particular echo sounding needs. A transducer is a ceramic element encapsulated in a suitable housing. When a voltage pulse is applied to the element, it will vibrate at its designed frequency. The vibration is called a piezoelectric oscillation, and it can be made to recur many hundreds of times per minute through the application of alternating voltage between two metal plates that sandwich the crystal. In the majority of depthsounders the quartz oscillates at a frequency of 200 kHz, or 200,000 cycles per second, well above the range of human hearing. Depthsounders that need to send a signal extremely deep will use a lower frequency crystal, usually at 50 kHz (50,000 cycles).

These sonic waves are aimed downward, reflect off the bottom or from fish, and then reenter the underwater transducer assembly, which picks up the weak return signals between transmitted pulses. The incoming echo strikes the element (also called a "plate" or "crystal") and applies a tiny mechanical stress that travels to its opposite face; the crystal is set vibrating mechanically for an instant, and in doing so generates an alternating voltage between its two faces. The vibrating sound pulses are thus turned back into electrical energy that is then fed through the metal plates and back to the receiver amplifier. Shallow water sounders may generate up to 25 watts of peak power (from the transmitter amplifier) so that the sound pulse will strike the bottom with enough force to return a substantial echo. In deepwater units, over 100 watts of power must be used to pick up the distant bottom, this

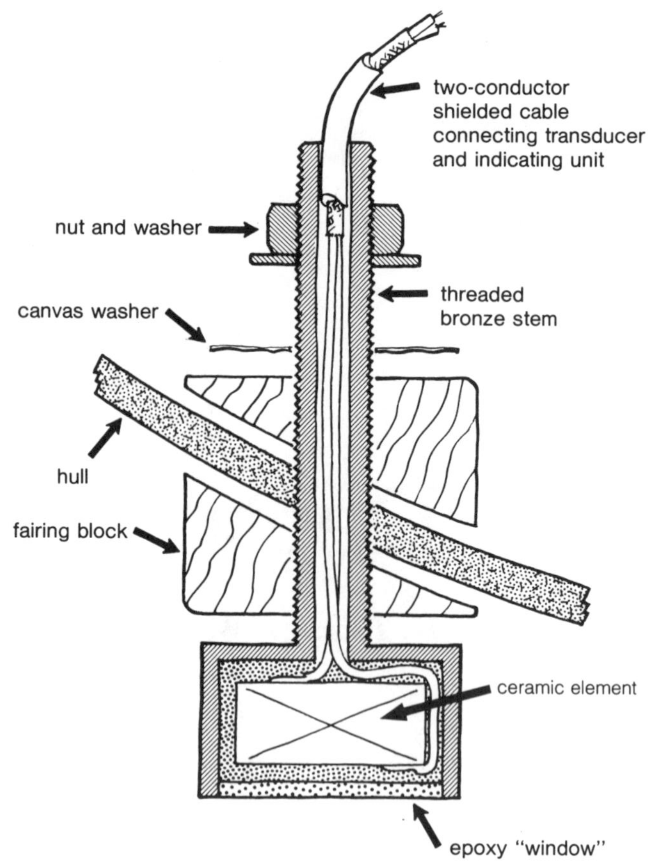

Cross section of a through-the-hull transducer mounting, showing the fairing blocks that must be hand-carved to fit the hull shape precisely while keeping the transducer in a vertical position.

power applying more than 400 peak volts to the transducer's piezoelectric mechanism.

The pulse repetition rate will vary between and within echo sounders depending on how deep they are intended to sound. On shallow sounding units, pulses are transmitted and received up to 1,500 times per minute. On deeper sounding units, more time is needed for the echo to be returned, so the pulse repetition rate may be as low as 60 times per minute. This lets the receiver section listen for the weak echo before the transmitter section fires off another ping. Shallow sounders will also transmit short pulses, and deeper sounders will transmit longer pulse lengths. For an analogy, imagine yourself in a boat in the fog, giving blasts on a foghorn and listening for echoes from a distant shore. Because the sound may be scattered by the shore and the fog, and intermittent background noises may intrude, you'll stand a better chance of picking up the echo of a longer blast. On most sounders offering more than one depth range to select from, pulse length and repetition rate are both set automatically when you "dial" the depth.

A 200 kHz transducer can be used for fishing, navigating, or both. Like transducers of other frequencies, it sends out a cone of acoustical energy that gets wider as it goes deeper, the usual beam width at 200 kHz being 9 to 15 degrees. (Beam widths up to 45 degrees are available for shallow water fishing.) With a 9-degree beam, a 200 kHz transducer is effective at depths up to 600 feet; at 100 feet, it will "see" a circle about 14 feet in diameter, giving quite a detailed picture of the ocean floor. If fish are right under you— even individual fish of moderate size—the 200 kHz, 9-degree transducer will pick them up quite nicely. Most digital indicators (which are used strictly for piloting) and flashers (which have some fishfinding capability) use 100 to 200 kHz transducers.

For a larger picture of the ocean floor, you need a physically larger, lower frequency transducer. The 50 kHz transducer may have a beam width of up to 54 degrees, covering a 100-foot-diameter circle in 100 feet of water. This lets you see fish activity off either side of your boat, and the lower frequency enables you to sound much deeper bottoms (3,000 feet or more, depending on the transmitter power). You lose a bit of definition, however; you won't be able to pick out those small rocky protrusions on the ocean floor, as you can with a tighter beam angle, and small fish may escape detection. For somewhat improved resolution at depths beyond the reach of a 200 kHz sounder, a 50 kHz transducer with a narrower beam width (as narrow as 16 degrees) might be a better choice. Regardless of beam angle, however, no 50 kHz transducer will provide the shallow water resolution of a 200 kHz transducer. The 50 kHz transducer is commonly used where both navigation and deepwater fishfinding capabilities are desired.

If you want the best of both worlds, choose a depthsounder with two transducers—perhaps a 200 kHz transducer with a wide cone angle for shallow water fishing, and a 50 kHz transducer with a narrow cone angle for deepwater fishing and

sounding. You can use a transducer switch box to select between them. Fishfinders frequently use this configuration.

Check with the local fishing community to see what type of transducer they are using for a particular fishing area. If you already own a depthsounder, you can switch transducers providing you match the new transducer frequency and impedance with the frequency and power of your depthsounder transmitter. Check your instruction manual for the operating frequency of your sounder. The converse is also true— you can replace your indicator unit with a "smarter" or more sophisticated unit (from a different manufacturer if you desire), yet keep the old transducer. Again, you must ensure that the frequency and power of the depthsounder transmitter matches the transducer. If the transducer is physically small in size, it could be damaged by an extremely high-powered transmitter. High-power, low-frequency transmitters, such as those used for extremely deep sounding, require massive transducer systems and will not be compatible with the typical 200 kHz transducer. With this exception, however, mixing and matching components is quite feasible and certainly advantageous, since transducers seldom wear out. You may not get a discount on the indicator unit by telling the seller you don't want the new transducer, but at least you'll have a spare transducer, and you'll save the hassle and expense of another transducer installation.

Transducer Installation

The transducer must be in contact with the water for best performance, and proper location and fairing of the installation is critical. On larger vessels, power or sail, the transducer is usually mounted through the hull. Although manufacturers' instruction manuals for other types of marine electronic equipment usually offer complete installation directions, this is often not true of depthsounder transducers mounted through the hull, because the manufacturers assume, wisely, that you will leave the job to a boatyard. If you do decide to go it alone, however, the key to good depthsounder performance is to find a spot where no aeration will disturb bottom soundings.

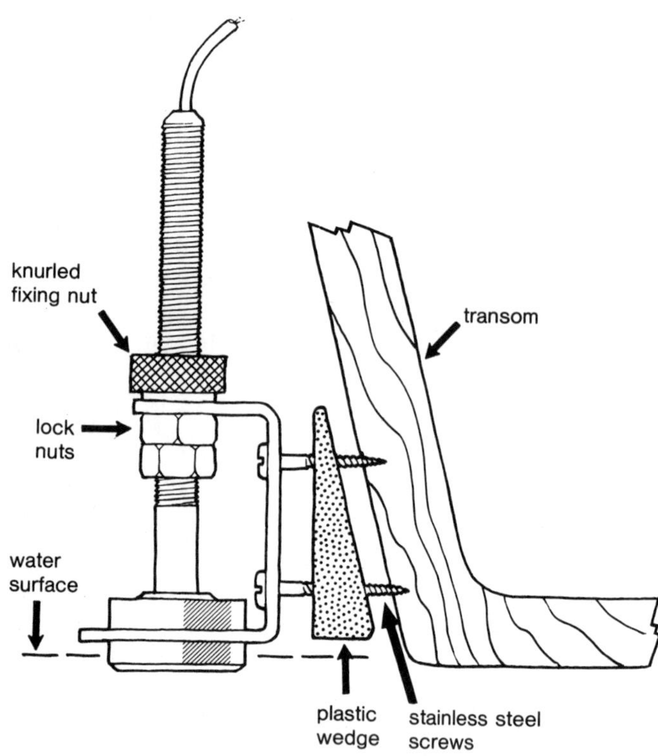

Transom-mounted transducers are subject to erratic operation caused by air bubbles flowing off the boat's bottom. They are practical for slow speed boats where the transducer is totally submerged at all times, but on planing hulls they may provide only intermittent information. Some brackets are gimballed, and by pivoting the transducer it is sometimes possible to find a position that will minimize erratic readings.

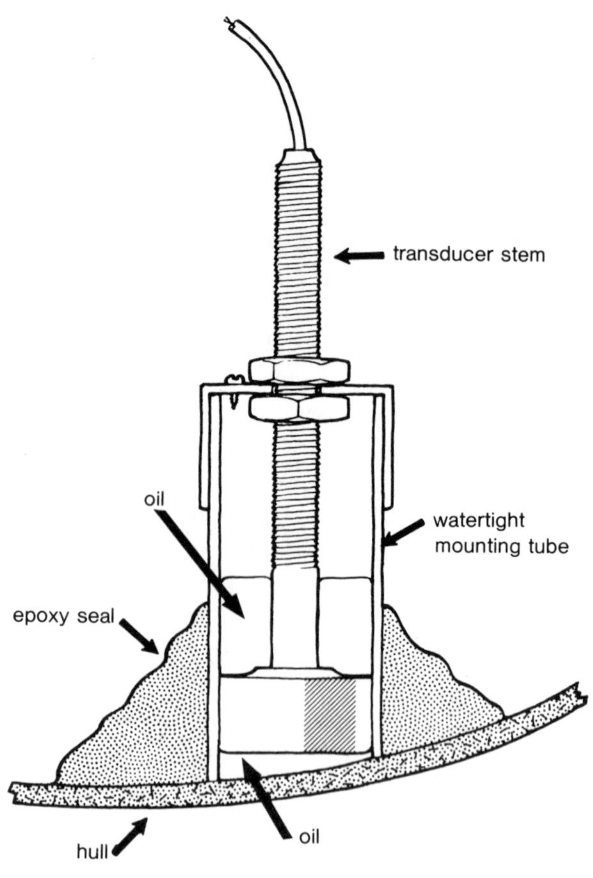

There are various means of mounting a transducer inside the hull, but almost always, the installation must be custom fitted. The transducer must be positioned vertically and the face of it submerged in oil to exclude any air between it and the hull. Any transducer with a threaded stem can be installed in such a watertight compartment.

The placement in a sailboat must be far enough from the keel that signals are not deflected when the boat is heeled. To get around this you can mount a transducer either side of the keel with a gravity switch automatically selecting between them. Many available depthsounders can be modified for this system.

The transducer is mounted through the hull using a special fairing block that will keep it pointed directly down toward the ocean floor when the boat is on an even keel. These are streamlined to minimize drag and to shield the transducer from turbulence, bubbles, and damage. Sounders that are professionally placed through the hull usually give superb performance, and if you are into seri-ous long-distance cruising, this is the only way to go. But let a professional (often the boatyard) do the mounting unless you're sure you know what you're doing. Transducers *can* be mounted by a scuba diver with a minimum amount of water let into the bilges, but if you want to try this trick, get an experienced diver. It would be much better to have the transducer mounted during your next haulout.

Most manufacturers supply a transom bracket for small boats that allows you to mount the trans-ducer off the stern (but still underwater), and some such brackets allow the transducer to self-adjust its orientation depending on your speed through the water. A slight miscalculation of transducer placement can seriously compromise readings even at moderate speeds, however, and on planing hulls the readings may be intermittent.

On some boats, you can mount the transducer inside the hull and sound through the wood or fiberglass. The accuracy of readings will not be affected, but the maximum sounding depth will be reduced by as much as 75 percent or more, and the ability to find fish may be all but eliminated. This system will not work at all with most cored hull constructions or when the laminate has local voids.

If you decide to try an inside-the-hull mounting of the transducer, follow the manufacturer's instructions precisely. The usual first step is to try out the transducer at different points below the waterline. The face of the transducer must be flat against the wood or fiberglass hull and covered with water or oil. Putting your boat in fairly deep water, move the transducer around in the bilge and find a spot where you get a solid reading of the bottom. If you find the results satisfactory, you can spring for one of the commercially available kits designed specifically to mount any type of transducer to the inside of your hull.

One further note: the marine electronics specialty dealers do many installations, but they will *not* install transducers. This job is usually done at a boatyard.

The transducer assembly is linked to the depthsounder equipment by a two-conductor coaxial-type cable. This cable should never be nicked or cut. Allowing moisture to enter a crack in the outside black sheath could damage the entire affair. Since this cable is carrying both transmitted and received signals, it must be shielded from external noise. *Never* route it beside the primary wires in your ignition system, and while you *can* run the cable beside other electrical wiring for short runs, it's best to keep it as far away from them as possible. Here are some examples of wires you want to avoid:

- Wires to fluorescent lights
- High-voltage wires from your SSB radio telephone antenna circuit
- Satellite antenna wires
- Bait tank wiring

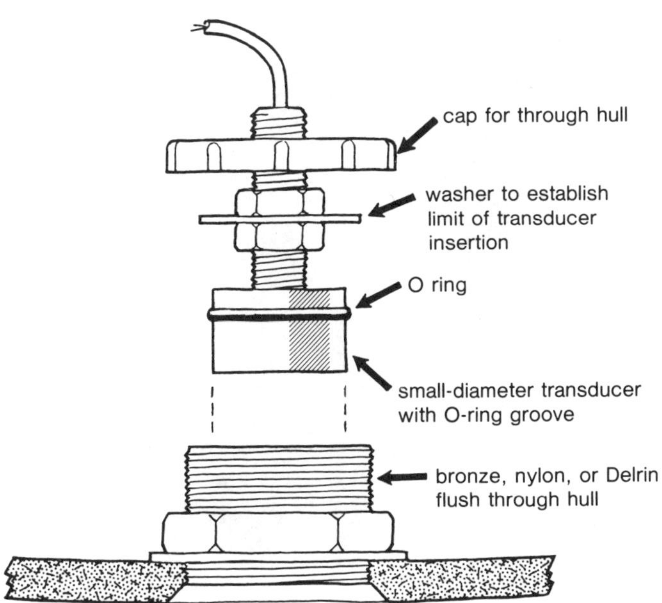

In waters where there is a heavy growth of barnacles, it is sometimes necessary to remove the growth from the face of the transducer to keep it operating properly. Special through-hull fittings are available that permit withdrawal of the transducer into the bilge for cleaning. One must screw a cap over the fitting when the transducer has been removed. A small amount of water will rush into the hull during the moments the transducer is being withdrawn and before the cap is screwed down.

The transducer cable usually contains an outside braid to keep static out and the signal in. Should the cable be severed, you must carefully reattach all of the wires, making sure that the braid covers up the entire works.

The end of the transducer cable simply plugs into the back of your depthsounder. Watch this connection! If neglected, it may become loose, corroded, and then intermittent. Make sure also that none of the wires pull free from the chrome connector collar. Intermittent bottom readings at slow speeds are sometimes the result of a loose or poorly connected transducer plug.

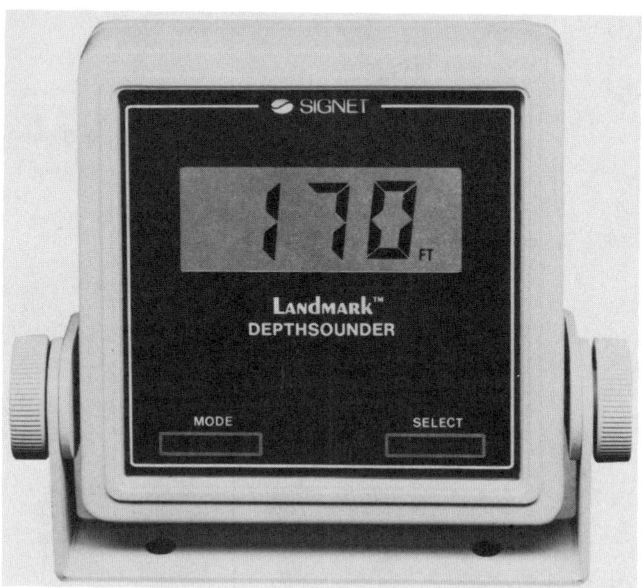

An inexpensive digital sounder that does one thing and does it well.

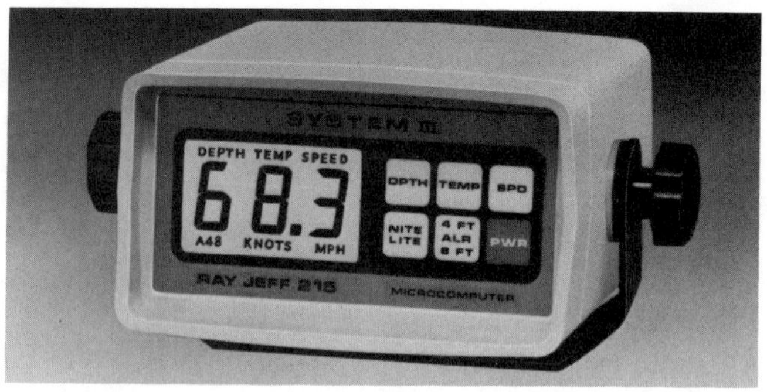

This indicator is matched to a "triducer," a through-hull pod housing an impeller and a temperature sensor in addition to a depth transducer. This seems to have been accomplished without noticeably affecting the efficiency of the individual sensors.

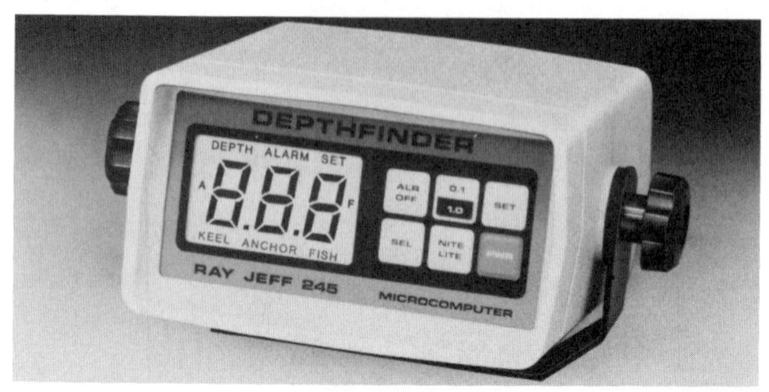

Despite advertising copy to the contrary, digital depthsounders are not good for fishfinding. The audible alarm which, in a few such units, is supposed to indicate fish, merely indicates the presence of any target—debris, fish, weeds, etc.—between the boat and the bottom.

Indicator Units

Some manufacturers refer to their depthsounders as "sonars," but this is really not accurate. A true sonar is a device that sends out energy horizontal to the surface of the water, while a depthsounder, of course, directs its energy perpendicular to the water surface. Sonars are very expensive pieces of equipment designed for commercial fishing applications—spotting large schools of fish, rocks, or obstructions on either side, astern, and ahead of the vessel. Depthsounders won't do this. Even if you turn a depthsounder's transducer horizontal to the water, it won't do it, because it does not possess the right type of transducer and associated circuitry to scan horizontally without picking up the water surface as a false target.

Manufacturers have developed extremely powerful transmitters and extremely sensitive receivers to be housed in the indicator unit, but don't be overly influenced by advertised transmitter powers. Since these claims may refer to peak watts, peak-to-peak watts (a highly inflated figure), average watts, or useful watts, comparisons are difficult. For better or for worse, the better indication of the performance of a sounder is its price tag—the higher the

price tag, the more watts in the transmitter, and generally the greater the sensitivity—that is, the ability to see weak echoes in the receiver.

The difference between a depthsounder and a fishfinder lies in the indicator unit and the microprocessor circuitry associated with a particular type of display.

Digital readout indicators are best for anchoring and handy for piloting, and they are the frequent choice of cruising and racing sailors and other consumers whose concern is depth in feet, fathoms, or meters, not where the fish are or what

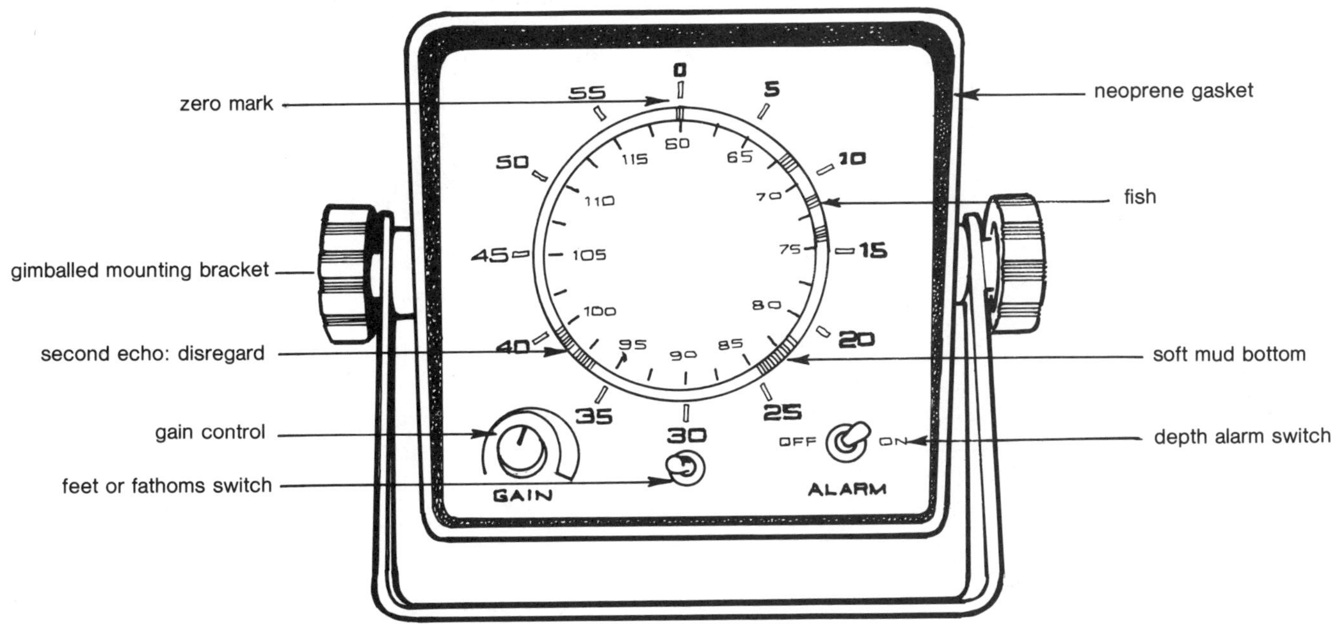

A flashing indicator is inexpensive and provides both bottom and fishfinding information. The price range with transducer is $150 to $450, and a typical size is 8¹/₂ inches high, 6¹/₂ inches wide, and 4 inches deep.

the bottom is like. Manufacturers are getting away from duller, light-emitting diodes (LEDs) to the more brilliant liquid crystal display (LCD) readouts, which can be seen easily in direct sunlight and are backlighted for night viewing. The LCD readout also draws less current, so your depth indicator can operate off portable dry cell batteries. These readouts are reliable and efficient and are now considered "state of the art." Light-emitting diodes are fast disappearing.

Digital sounders and meter-movement (also called "analog") sounders are both easy to read. In addition, digitals have no motors or moving parts, so they operate silently and effectively. Some digitals sport innovative features such as depth alarms and anchor alarms, products of microprocessor technology. You need only predetermine at what depth or depths you wish the alarm to sound. If you aren't watching the unit and get too close inshore, your alarm will let you know immediately. If you're asleep at night and are blown off- or onshore, the alarm will sound off letting you know you have exceeded a minimum or maximum preset depth setting. Sounders offering this feature differ widely in the available range of depth alarm settings. "Forward looking" sounders will even keep track of the rate of change in depth, giving a warning alarm when the water shoals rapidly toward a critical depth. Digital indicators with averaging circuits are advertised, and these will ignore wildly errant, ephemeral readings caused by debris or bubbles, or by a powerboat's hull momentarily lifting from the water at a high speed. Some digital sounders will automatically adjust raw depth readings (which, by definition, indicate depth below the transducer) to depth below the keel or below the water's surface. Some units are wired to speakers and programmed to deliver vocal depth reports.

Flashing indicators or flashers utilize a neon bulb or LED mounted on a rotating disk that is driven by a constant-speed motor. On each rotation a switch is magnetically activated, causing the transmitter to send an electrical pulse to the transducer. When the weak pulse from the returning echo is amplified, the bulb or LED is lit up and displayed against a circular calibrated scale. Be-

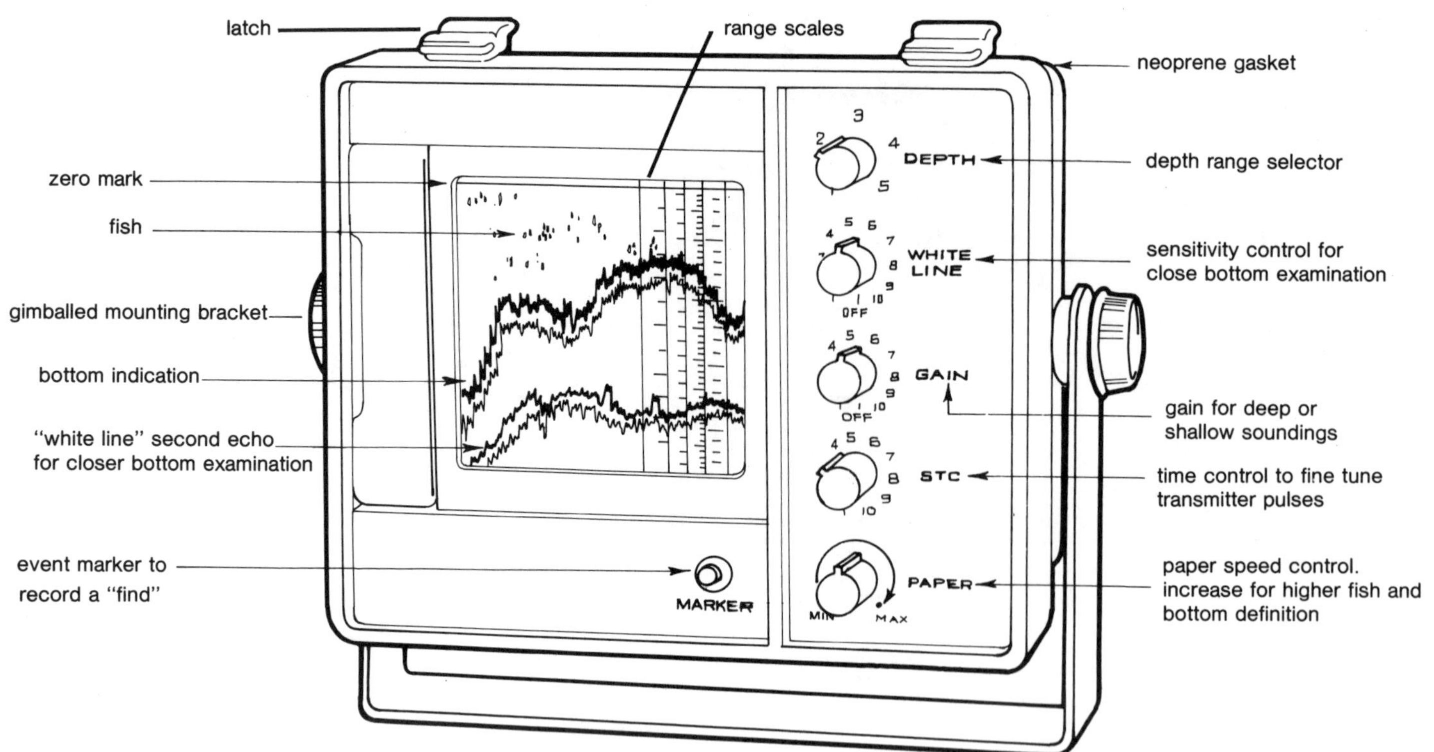

A chart recorder. This type of sounder is a favorite among fishermen, offering a tremendous amount of bottom and fish information. It is extremely useful in navigation as well as fishfinding. The price range with transducer(s) is $300 to $1,000, and a typical size is 9½ inches high, 12 inches wide, and 4 inches deep.

cause the disk rotates at a constant rate, the arc swept by the bulb in the time between outgoing and returning pulses is directly related to the depth as read off the scale. The disk rotates so rapidly that the flashes appear motionless but for changes in depth. Random flashes usually indicate fish, and if you watch a flasher closely, you can see the return echoes of even one or two large fish just off the bottom. Flashers are sometimes used by sportfishermen because they are inexpensive and they work well providing you keep your eye on the dial. Most flashers give you a number of depth ranges to select from. In making a selection you are in actuality adjusting the speed of the disk's rotation, with slower speeds being required for greater depths.

Flashers will also indicate the type of bottom below you. If you receive a broad bottom flash, it's probably mud. A sharp bottom flash is likely to be hard sand or rocks, and an irregular bottom would probably show up as a very broad bottom flash that changes in width.

The latest innovation in flashers is a no-moving-parts liquid crystal display. The round dial remains, but the LCD black lines take over for the rotating disk and its mounted display lamp. The same LCD display can also give you (with separate sensors) water temperature and boat speed; digital depth displays and predetermined depth alarms are possible, too. Watch the LCD rotary-type display grow in popularity.

A flasher should have a gain control knob on the front panel with which to adjust the sensitivity of the receiver circuitry. Tuning this efficiently takes practice. Too much gain causes loss of target discrimination and may even cause false targets to appear; too little gain will cause weakening or loss of bottom echoes.

If you are really into serious fishfinding, you'll want a recording depthsounder, also known as a "fishfinder." There are two types—**chart recorders** and **video** or **CRT** (cathode ray tube) **recorders**. Chart recorders are excellent, and they are quite useful in navigating in deep or shallow waters. The chart paper will give you an accurate, permanent record of the bottom and everything in the water column. A typical reading might depict your downrigger being brought up from 70 feet for a lure change, large fish feeding off the bottom, game fish just below the surface of the water, and a well-defined wreck or obstruction on the ocean floor that might be a great candidate for skin diving exploration. Chart recorders will also assist you in following charted bottom contours to find your way home in the fog, and they will let you see the trend of a shoaling channel as you reenter a harbor.

Chart or graph recorders are also getting smarter, so if you're going to get a graph recorder, get a smart one. You can now adjust a dial to zoom in on the exact depth you want to examine. Want a closer look at that possible game fish at 40 fathoms or the ocean floor beneath you? Zoom in and take a long gander. And if you need to mark a spot on the chart where your downrigger hit a snag, push a button and there it is!

Many recorders feature bottom definition circuitry, variously referred to as "white line," "gray line," or by one of several other monikers. By purposely reducing bottom echo, this circuitry isolates fish that are floating just off the bottom. Some depth recorders will allow you to change paper speed, too; a faster paper speed yields better bottom definition, "definition" implying the ability to detect small targets and deep bottoms (sensitivity) and to separate crowded targets (resolution). All of this is available in the more sophisticated depth recorders that offer variable paper speed, variable pulse length, and a combination of switchable transducers in addition to power and gain controls.

The latest innovation in depth recorders is the liquid crystal display, in which moving chart paper is replaced by black markings on an LCD display screen. These units are usually completely waterproof, and they'll let you see up to 32,000 bits of information at one time. You can recall one full screen of past information by pressing a single button, and an auxiliary tape recorder will allow you to record the digital information for later replay. The LCD readout is just as detailed as your digital watch, so you can imagine the precision of each target. The only characteristic you must get accustomed to is that every target is a miniature square, as opposed to a faithful reproduction of what the

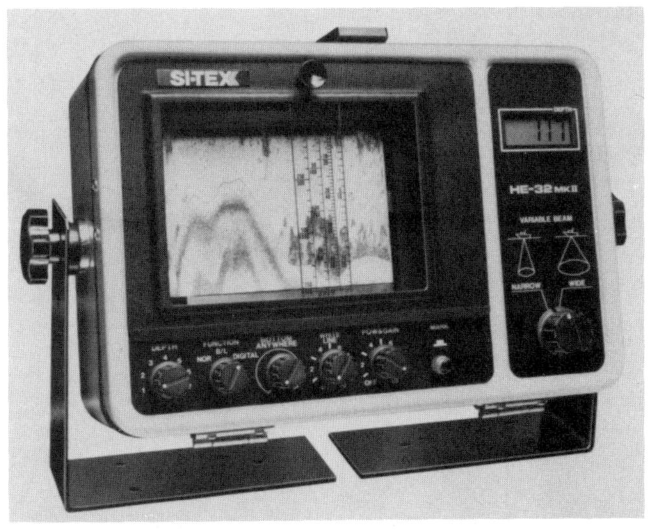

This chart recorder is matched to two transducers contained in one pod, so a narrow or wide cone angle can be selected as desired.

transducer is "seeing." An ascending bottom line that would appear smooth on a chart recorder will look like miniature ascending steps on the LCD readout. We should see better resolution in the future, as manufacturers put dots in every LCD frame.

A new feature of LCD graph recorders is the multicolor pixel display. Recent innovations in LCD readouts now allow manufacturers to develop colors that represent different echo intensities. The LCD color display will indicate extremely strong echoes as red pixels, while echoes with slightly less strength may show up as yellow or green. Some manufacturers claim that certain colors indicate given species of fish, but this is a little hard to swallow. Echoes return quite as nicely off kelp, for example, as off fish between the surface of the water and the bottom.

A drawback to the color LCD recorder is its requirement of direct viewing and the right amount of light for best color discrimination. If you don't have ideal light, or if you are positioned a little to the left or right, the colors are very hard to discern. Take a close look in the showroom before you get caught up in the enthusiasm for color LCD recorders.

Video sounders represent another area of rapid development within the depthsounder field. Actu-

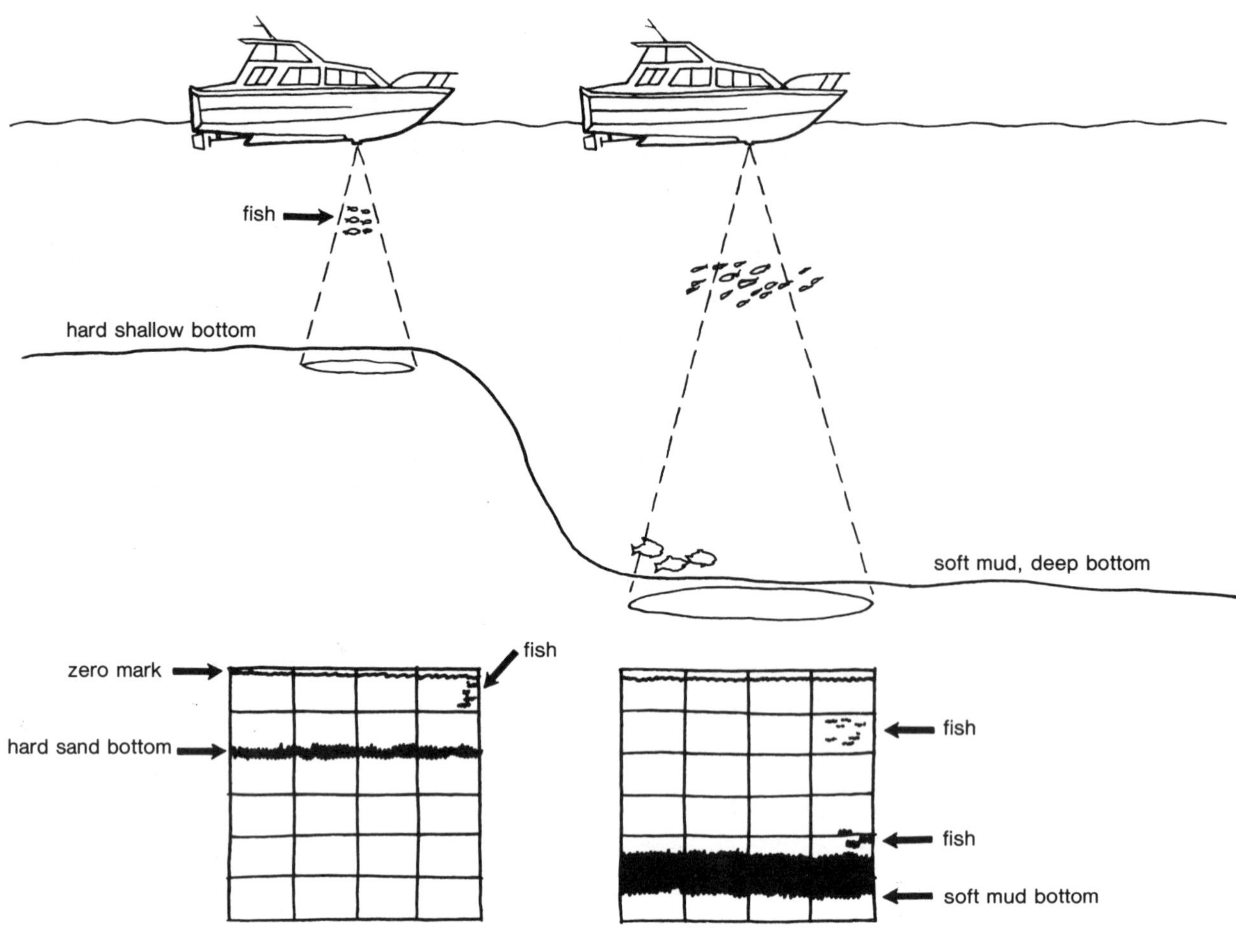

Vertical lines on a recording sounder's display represent time that has elapsed as the display moves from right to left, and the horizontal lines represent depths in feet, fathoms, or meters.

ally, monochromatic (black and white) and color video are not new to commercial fishermen. Video sounders and video sonar have been part of their fishing arsenal for years. Only since 1985, however, has digital computer technology allowed this equipment to be manufactured smaller, more powerful, and at lower prices. One commercial fisherman of my acquaintance, operating out of San Pedro, California, has been using color video for 10 years and claims to be able to identify fish species by their echo colors. Years of experience in his home waters have given him this ability; in different waters, with different species of fish, he would have to teach himself all over again by consultation and by trial and error.

The new megabyte computers inside video sounders will catalog returning echoes by their in-

tensity, the relative sizes of the targets, and their depths. Thousands of bits of information can be processed and then presented on a 10-inch screen. The first low-cost video sounders were monochromatic—black and white, amber and white, or gray and black. These are amazing machines because of their wide dynamic range, enabling them simultaneously to pick out weak echoes close to the bottom and strong echoes near the water surface. This is a feat beyond the capability of a chart recorder, on which the gain would need to be set so high to pick up a tiny bottom target that anything near the surface would be obliterated.

The monochromes sell for between $500 and $1,000. Multiple levels of quantitization allow these hardworking units to elicit five to nine different shades of intensity to differentiate among large

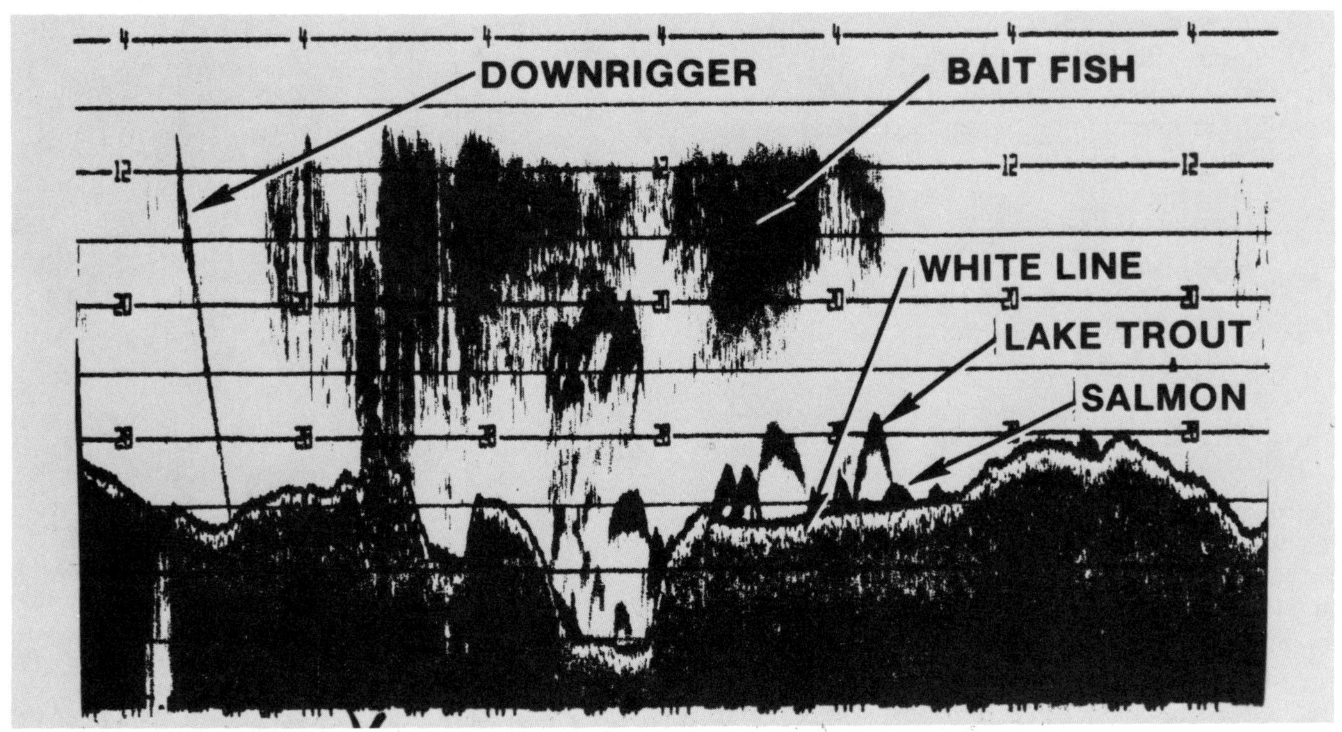

A sample chart paper showing target interpretations.
(Courtesy Ray Jefferson)

and small targets (strong and weak echoes). Just as there are still black and white televisions and viewers who swear by them, monochromatic videos will not be completely superseded by the smarter, color videos that are taking their place aboard the boats of serious fishermen.

The color videos usually incorporate a 10-inch television-type display. Here is an example of the display colors and what they may represent:

Color	Meaning
Red	100% relative echo return
Orange	85% relative echo return
Yellow	70% relative echo return
Dark Green	55% relative echo return
Light Green	40% relative echo return
White	25% relative echo return
Light Blue	10% relative echo return
Dark Blue	No echo return

It takes a day or so to begin to figure out what's going on below the water and what you are actually seeing as color-coded information. A hard, rocky bottom will show up red, while a bottom with kelp or mud might show up as red overlain by orange. Large fish with big air sacs will cause strong echo returns and may show up as orange, yellow, or green targets. A school of fish or shrimp rising off the bottom might show up as green or white. Every species of fish has its own color characteristics, and once you learn your machine, you can interpret what's down there with good accuracy.

The dynamic range and sensitivity of a color video is so sharp it may also show up thermoclines (density stratifications in the water caused by sharp variations in temperature). Since certain fish like to feed in thermoclines, the video sounder adds one more fishfinding tool to the electronic tacklebox.

As in a color television set, the images on the screen of a video sounder are well defined at a distance. At close range, however, rows of tiny dots (usual resolution 256 x 256 dots across the screen) are visible. The color sets can change the color of an individual dot to any one of 8 to 16 hues. The following display controls may be available on a color video:

- Depth range
- Zoom minimum and maximum
- Screen speed indication
- Loran information (when interfaced)
- Boat speed
- Speed log
- Water temperature
- Digital bottom depth
- Depth in feet, fathoms, or meters
- Noise rejection indication
- Percentage of gain
- Alarm modes
- Variable range marker line

The transducer offers discrete shallow and deep water frequencies, a speed sensor, and a water temperature sensor in an assembly that is hardly larger than older, conventional transducers. (It is possible to mate your existing transducer to a video sounder if you can find a video indicator having the same frequency and power as the indicator you're replacing. In this instance, however, resolution will probably be better if you buy the transducer recommended by the video manufacturer.)

The digital logic of the video sounders allows you to freeze pictures, store up to ten displays in memory and later compare them with a new display, split the screen, or lock on the bottom. If you are sounding at 400 feet and spot something just off the bottom, you can zoom into the bottom area—perhaps with the 380-foot contour at the top of the screen and 410 feet at the bottom of the screen. Targets within those depths are then enhanced, and you get a close-in look at the ocean floor.

Another feature is an audible fish alarm. Set your desired maximum or minimum depth range, turn the alarm on, and wait for it to be triggered, indicating yellowtail at the selected depth.

Videos are best operated in the shade, so you can see what's happening on the screen. Sun can easily wipe out the colors when a screen is viewed on a bright flying bridge. These units will also create some video interference to a loran set if not properly grounded. Their horizontal sweep can easily knock out weak loran signals, so make sure the video installation is well grounded and that all wires are properly shielded.

If you are already equipped with a video sounder, you may wish to add a $260 device called

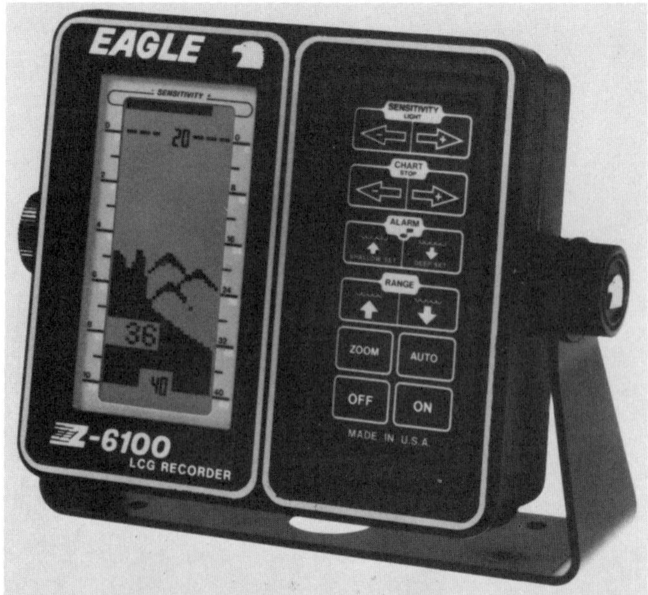
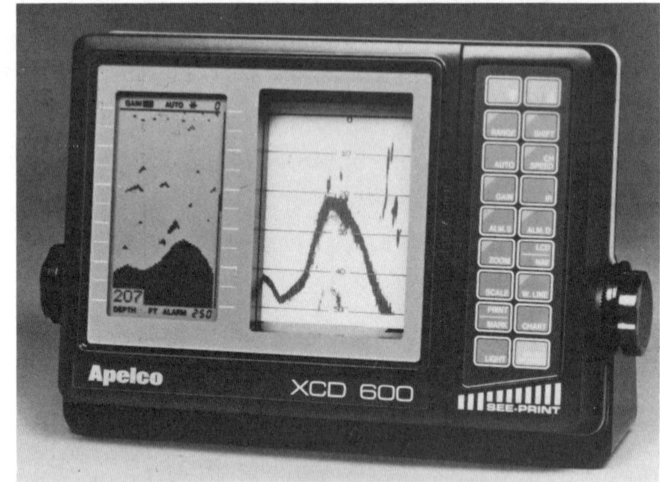

Two LCD (liquid crystal display) recorders, one having a paper chart recorder as well. Resolution of the readouts is a function of the maximum depth setting and the number of dots in the screen, and is improving with newer models.

"Depth Talker," which simply plugs in between your transducer and the video display unit. This device won't affect the video readout at all, but it will call out depths less than 100 feet in plain-enough English. The idea is that you won't have to study the screen to see what the depth is before anchoring. Just listen for the spoken word. Although there's no change in tonal inflection when a

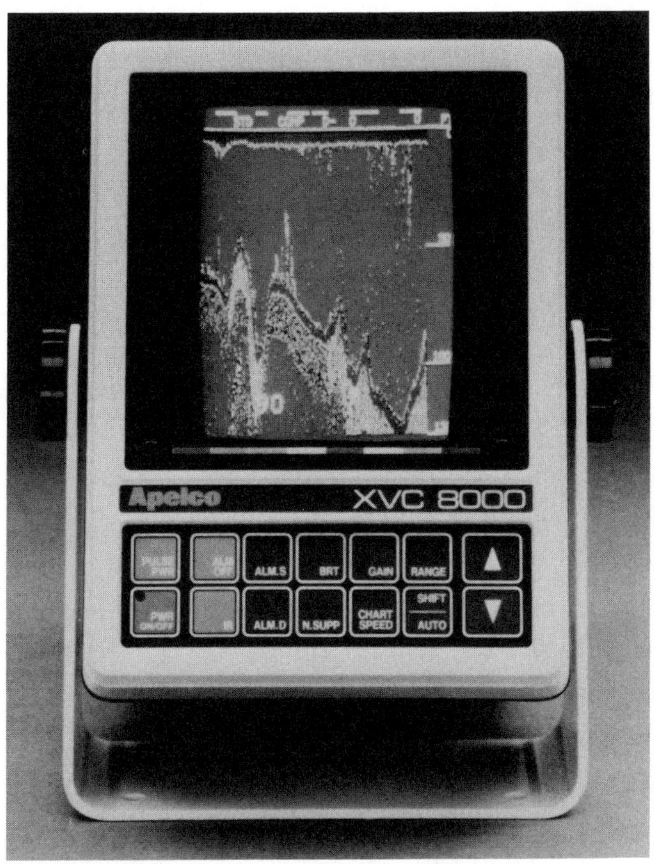

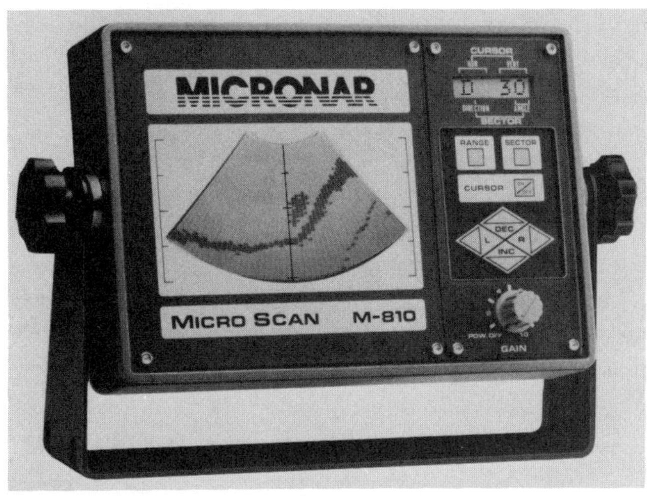

This is the only side-scanning sonar available for less than $1,000. Unlike a depthsounder-fishfinder, this sonar can search for targets to the left or right of the boat, a feature likely to be of interest to serious fishermen.

"Color video fishfinders," as the industry terms them, offer much better resolution in their cathode ray tube displays than one can obtain with LCD recorders, particularly in deep water. Although some LCDs utilize colored displays, the use of colors and various intensities to distinguish among targets is more refined with video sounders. These fishfinders are sometimes inaccurately referred to as sonars in advertising copy, but they lack sonar's ability to scan the waters surrounding a boat. Like any depth-sounder, video sounders "see" only within the limits of the transducer cone below. Prices range from $700 to $8,000 or more.

boat rapidly approaches grounding (no trace of panic!), the unit is useful when you've got your eyes on the bow and want to know your depth in feet or fathoms. For more information, write Bob Tendler, Depth Talker, Inc., 19 Lawrence Avenue, Chestnut Hill, Massachusetts 02167.

Before you purchase a depthsounder, consider whether or not you need one that is waterproof, or at least weatherproof. Manufacturers are finally giving us completely waterproof sounders that can be hosed down after a day of fishing, and if you're going to be running your depthsounder in an open cockpit, then by all means select a unit that will take this type of punishment.

Installation of the indicator unit is straightforward, and you can accomplish it yourself. The units don't draw much current, and manufacturers usually suggest No. 10 wire for the positive and negative leads. This wire for voltage goes over to a switch panel, and there is nothing tricky about it— red to hot, and black to negative. Most sounders don't require an additional wire to ship's ground, but it's a good idea, helping to reduce static. A single No. 12 wire from the ground post on the sounder (or under any screw) should be run to your ship's ground plate or any common ground. Ground to your engine if nothing else is available.

Preventative maintenance involves cleaning the power connection, the transducer plug, and other connections regularly, and making sure the unit is firmly in its mounting bracket. There is little to do inside the unit; sounders seldom require recalibration.

It is recommended that you paint the face of the transducer with antifouling paint. Usually one thin coat of bottom paint will keep the barnacles from forming on your transducer. Go with the bottom paint, not the barnacles. (If you should ever encounter growth of any kind on your transducer, take great care in removing it. That quartz crystal is protected only by a thin layer of plastic, and if the plastic gets gouged it skews the acoustical energy in different directions.)

Depthsounder malfunctions are usually traceable to the indicator unit, not the transducer, and the unit can easily be removed and sent back to the dealer or manufacturer for repair. Rarely do well-installed transducers go bad unless they have been physically damaged. If you treat your equipment properly, your depthsounder should work flawlessly for years.

Finally, let's talk about something you might not have known a sounder is capable of—and that's serious navigation. For example, you can tell what type of bottom you are over from a recorder, video recorder, or flashing indicator, though not from a digital. Over a sandy bottom, you will see a hard bottom indication, while kelp on the bottom will appear as a broad bottom signal. If you are cruising in unknown waters and following charted depth contours, remember that charts are based on mean low water. You can safely navigate around islands by following bottom contours and watching the chart. Unsure what island you are going around? Look for underwater hills and valleys by cruising in a straight line alongside the island, and match the depth configurations you see to the chart. In similar fashion, with a chart, dividers, a stopwatch, and a little mathematics, it's possible to use a depthsounder for determining speed over the bottom. Watch those bottom contours—charts are quite detailed for this purpose. Your charted depth contour can also constitute a line of position to substantiate bearings from a radio direction finding station.

Buying Tips

The information contained in this chapter may be used as a springboard for the examination of manufacturers' literature and advertising claims. By now, you probably have a good idea as to what kind of depthsounder best meets your particular needs. If not, the accompanying chart may help you decide. In addition, a complete presentation of manufacturers, specifications, and prices appears in Appendix C. When you have sifted through all of these, your chief remaining question is likely to be where to buy the unit.

A depthsounder is a logical piece of equipment to purchase from a catalog-discount house, because you're not giving up much in the way of service in order to save some money. Marine electronics dealers will install the indicator unit in your boat, but that part of the installation you can easily do yourself. They will *not* install the transducer, so you'll have to pay a boatyard to do that or struggle through it yourself in any case. Finally, if the indicator unit does malfunction, it will be easy to disconnect and ship directly to the manufacturer for repair.

TYPE OF DEPTHSOUNDER AND SUGGESTED USE

Type of Depthsounder	Navigation in Small Boats	Sport Fishing	Large Motor Yachts	Trailerboats	Sailboat Cruising	Sailboat Racing	Fresh-water Fisherman	Powerboat Racing (high performance)	Commercial Fishing
Digital	✓		✓	✓	✓	✓		✓	
Flasher	✓	✓							
Paper Chart Recorder		✓	✓		✓				✓
LCD Graph		✓		✓			✓		
Video, monochromatic		✓			✓				
Video, color		✓	✓						✓
Sonar, side scanning (less expensive unit)		✓			✓				
Sonar, color (full set)			✓						✓

3

VHF Radio

by Gordon West

Marine VHF radio enables you to communicate with other vessels up to 20 miles or more away, or with shore stations at greater distances. The VHF signals, having a short wavelength, are not much affected by local weather conditions, so VHF radio is a very reliable mode of marine communication. Before any other marine radio can legally be carried in your vessel, you must have a marine VHF transceiver aboard.

Virtually all VHF radios are manufactured in the Orient. State-of-the-art sets are synthesized—that is, they employ a single crystal that can be split by means of factory-programmed microprocessor circuits to transmit and receive on any existing marine VHF channel. The more traditional crystal sets cannot cover all the channels without employing multiple crystals, a more ponderous and expensive approach. Crystal sets are still being manufactured, but these generally offer only 12- or 24-channel coverage. Some 55-channel synthesized sets cost less than 24-channel crystal sets.

Armed with a basic understanding of the available VHF channels and a few simple rules to follow when choosing an antenna and a transceiver, you can put together a VHF station that suits your boat and satisfies your requirements. When you do, remember that perhaps the biggest problem facing marine VHF radio is the abuse of broadcasting protocol by pleasureboat operators. Mariners should learn and adhere to the standard marine VHF procedures.

The Channels

The marine VHF (very high frequency) band is centered around 156 megahertz (MHz), between Channels 6 and 7 on your television set. It is a group of frequencies standardized throughout the world by international agreement of the World Administrative Radio Conference (WARC). The band encompasses approximately seventy-eight 25-kHz channels between 156.050 MHz (Channel 1) and 157.425 MHz (Channel 88A) for ship transmitting; certain public telephone shore stations will also transmit on frequencies from 161.800 MHz (Channel 24 receive) to 162.025 MHz (Channel 88 receive, not applicable in the United States). The VHF band also includes several national weather transmitting frequencies centered around 162.550 MHz (weather channel 1 receive). Those frequencies between the ship transmit and shore transmit channels are allocated to other radio services (157.450 MHz to 161.595 MHz).

Cellular Phones	850 MHz
Channel 7 (Television)	179 MHz
Very High Frequency (VHF) Radiotelephone	156.0–162.55 MHz
EPIRB (Class C)	156.8 MHz
EPIRB (Classes A and B)	121.5 MHz
Omni/VOR	108.0–118.0 MHz
FM Radio	88–108 MHz
Channels 2–6 (Television)	54–88 MHz
Citizen's Band (CB) Radiotelephone	26.9–27.3 MHz
High Frequency Single Sideband Radiotelephone	4.0–23.0 MHz
Medium Frequency Single Sideband Radiotelephone	1.6–3.5 MHz (1,600–3,500 kHz)
AM Radio	530–1,600 kHz
Marine Radio Beacons	190–400 kHz
Loran-C	100 kHz
Decca	70–130 kHz
Omega	10.2–13.6 kHz

The number of channels that a mariner has access to in the United States is actually no greater than 55, some channels being paired differently with shoreside channels in other countries. Examples are channels 23A and 88A. The letter "A" after any VHF channel indicates that the channel has a split capability, being used in the United States for simplex (same frequency) transmitting and receiving, while having another use elsewhere. Channel 23A might be used by Coast Guard auxiliary members (with special permission from local Coast Guard units) to communicate with Coast Guard vessels. Channel 88A is used in almost all areas of the United States for ship-to-ship simplex commercial communications as well as for fish spotting by aircraft. In other countries, such as Canada, Channel 23 (without the "A" designation) is used for communications to a shoreside marine telephone service where the transmit and receive frequencies are different (a duplex split of 4.6 MHz). In the Great Lakes, mariners use Channel 88 (no "A") in the duplex mode to place telephone calls with shore stations that transmit on a different frequency (4.6 MHz higher) than the ship stations.

Modern VHF sets have programmed microprocessor circuits that tell them on what channels to operate simplex and when to switch automatically to duplex (different receive). Since on the "A" channels (and only on the "A" channels), this may vary by country, sets are usually factory programmed for use either in the United States—designated as the "United States" or "domestic" mode—or elsewhere—the "foreign" mode. A few sets offer a manual override for switching modes. Apart from these few exceptions, no matter where you cruise throughout the world, all VHF channels are the same.

The most important channel is 156.800 MHz, Channel 16, the international distress and calling frequency. Every VHF radio must have this channel installed in it, and it must be monitored at all times when your set is turned on, except when you are operating on another frequency. The United States Coast Guard and other rescue agencies worldwide monitor this channel 24 hours a day, and it is also monitored by orbiting search and rescue satellites that will retransmit the distinctive sounds of an emergency position indicating radio beacon (EPIRB) signal, along with its approximate location, when one is received. This channel must only be used for distress and safety calls or establishing communications with another vessel before switching to a working channel. You are not allowed to talk more than 59 seconds on VHF Channel 16 unless you are making an emergency communication.

VHF Channel 6 is the safety channel, to be used only for intership communications relating directly to the safety of vessels. Use this channel to communicate priority or urgent traffic to another vessel. Every VHF set must have Channel 6.

There are six pleasureboat ship-to-ship communications channels: 9, 68, 69, 71, 72, and 78. Use any of these to communicate with other boats in your area. The messages you transmit must relate directly to the operation of your vessel, and all communications must be kept to a minimum, since other vessels in your area may wish to use the same frequencies. Yacht clubs, ship supply stores, and fuel docks may also use Channels 9, 68, 69, 71, and 78 to communicate with ships.

VHF Channel 70 (156.525), once available for voice communications between pleasure boats, is now reserved exclusively for digital selective signaling. Digital selective calling is new to the marine radio service but old technology to land services. The envisioned system would allow shore stations to digitally call silent ship receivers to alert them of

Channel Numbers	Type of Communication	Suggested Channel Selection		
		Recreational Vessels		Commercial Vessels
		6 ch.	12 ch.	12 chs.
16	DISTRESS, SAFETY & CALLING Intership & ship to coast	*	*	*
6	INTERSHIP SAFETY Intership. NOT to be used for non-safety intership communications	*	*	*
22A	Communications with U.S. Coast Guard ship, coast, or aircraft stations	1	1	1
1,5,65A, 66A,12, 73,14,74, 63,20,77	PORT OPERATIONS Intership & ship to coast		1	2
13,67	NAVIGATIONAL Intership & ship to coast		1	1
68,9,69, 71,78A,72	NON-COMMERCIAL Intership & ship to coast	1	3	
72	NON-COMMERCIAL Intership		2	
1,7A,9,10, 11,18A, 19A,79A, 80A,63	COMMERCIAL Intership & ship to coast			3
67,8,77, 88A	COMMERCIAL Intership			1
24,84,25, 85,26,86, 27,87,28	PUBLIC CORRESPONDENCE Ship to public coast	2	2	2
70	Intership & ship-to-coast selective signaling only. No voice communications.			
***88				
162.400 (WX-2) 162.550 (WX-1) 162.475 (WX-3)	NOAA WEATHER SERVICE Ship receive only	**	**	**

*These channels are REQUIRED to be installed in every ship station equipped with a VHF radio.

**The weather receive channels are half-channels (receive only) one or more of which are recommended to be installed in each ship station. Many manufacturers include one or more of these channels in their sets in addition to the normal six or twelve channel capacity.

***See limitations in table on page 26.

waiting messages. It might also be used to alert groups of stations that calls are holding. Commercial operators will benefit most from selective signaling. Why Channel 70 was pulled from the pleasure community and given over to commercially used selective signaling is not clear.

Pleasureboat operators are not allowed to use commercial frequencies, nor may commercial vessels use noncommercial frequencies. Of course, while it's illegal for pleasureboat operators to transmit on Channel 88A, for example, there is no harm in monitoring this channel to get the latest scoop from commercial fish spotters! Marine Channel 9 is shared by both commercial and pleasure boats, so if you wish to talk to a commercial fishing boat, first call him on 16 and then switch to 9.

When you hear the distinctive warbling sound of an emergency position indicating radio beacon on Channel 16, it means that someone in your area is in distress. The Class C EPIRB* will alternate automatically between Channel 16 and the adjacent Channel 15. EPIRBs are only one-way transmitters, so you won't be able to establish communications with the vessel in trouble. Your role is immediately to contact the Coast Guard on Channel 16 and let them know that you hear an EPIRB in your area. Call the Coast Guard when the EPIRB switches from Channel 16 to Channel 15 so you can hear their response. They may ask whether you have direction finding capability, and if so, what bearing the EPIRB is from your computed location.

The table lists several channels used for placing phone calls to marine telephone operators ashore. It's best to register ahead of time with your local marine operator before you attempt to place a phone call though their service. This is easily accomplished with a shoreside call. Marine electronics specialty houses may also have marine operator registration forms for specialized phone services not listed with your AT&T operator. Many of these offer terrific telephone service, including message handling and conference calling. A marine telephone service is a great bargain. There is no monthly fee if you don't use their service. You're only charged for the phone calls you place, and if

*Class C EPIRBs, the usual choice for coastwise vessels, transmit at 156.8 MHz. Class A and B EPIRBs for high-seas use transmit at 121.5 MHz.

| Channel Number | FREQUENCIES (MHz) | | CHANNEL USAGE |
	Ship Transmit	Ship Receive	Intended Use
1	156.050	156.050	PORT OPERATIONS AND COMMERCIAL (Intership and ship-to-coast). Available for use within the U.S.C.G. designated Vessel Traffic Services (VTS) area of New Orleans, and the Lower Mississippi River.
3	156.150	156.150	
63	156.175	156.175	
5	156.250	156.250	PORT OPERATIONS (Intership and ship-to-coast). Available for use within the U.S.C.G. Vessel Traffic Services radio protection areas of New Orleans and Houston.
6	156.300	156.300	INTERSHIP SAFETY. Required for all VHF-FM equipped vessels. For intership safety purposes and search and rescue (SAR) communications with ships and aircraft of the U.S. Coast Guard. Must not be used for non-safety communications.
7A	156.350	156.350	COMMERCIAL (Intership and ship-to-coast). A working channel for commercial vessels to fulfill a wide scope of business and operational needs.
8	156.400	156.400	COMMERCIAL (Intership). Same as channel 7A except limited to intership communications.
9	156.450	156.450	COMMERCIAL AND NON-COMMERCIAL (Intership and ship-to-coast). Some examples of use are communications with commercial marinas and public docks to obtain supplies or schedule repairs and contacting commercial vessels about matters of common concern.
10	156.500	156.500	COMMERCIAL (Intership and ship-to-coast). Same as channel 7A.
11	156.550	156.550	COMMERCIAL (Intership and ship-to-coast). Same as channel 7A. It should be noted, however, in certain ports channels 11, 12 and 14 are to be used selectively for the Vessel Traffic Service now being developed by the United States Coast Guard.
12	156.600	156.600	PORT OPERATIONS (Intership and ship-to-coast). Available to all vessels. This is a traffic advisory channel for use by agencies directing the movement of vessels in or near ports, locks, or waterways. Messages are restricted to the operational handling, movement and safety of ships and, in emergency, to the safety of persons. It should be noted, however, in certain ports 11, 12, and 14 are to be used selectively for the Vessel Traffic Service being developed by the United States Coast Guard.
13	156.650	156.650	NAVIGATIONAL (Ship's) bridge to (ship's) bridge. This channel is available to all vessels and is required on large passenger and commercial vessels (including many tugs). Use is limited to navigational communications such as in meeting and passing situations. Abbreviated operating procedures (call signs omitted) and 1 watt maximum power (except in certain special instances) are used on this channel for both calling and working. For recreational vessels, this channel should be used for listening to determine the intentions of large vessels. This is also the primary channel used at locks and bridges.
14	156.700	156.700	PORT OPERATIONS (Intership and ship-to-coast). Same as channel 12.
15		156.750	ENVIRONMENTAL (Receive only). A receive only channel used to broadcast environmental information to ships such as weather, sea conditions, time signals for navigation, notices to mariners, etc. Most of this information is also broadcast on the weather (WX) channels.
16	156.800	156.800	DISTRESS, SAFETY, AND CALLING (Intership and ship-to-coast). Required channel for all VHF-FM equipped vessels. Must be monitored at all times station is in operation (except when actually communicating on another channel). This channel is monitored, also, by the Coast Guard, public coast stations and many limited coast stations. Calls to other vessels are normally initiated on this channel. Then, except in an emergency, you must switch to a working channel.
17	156.850	156.850	STATE CONTROL. Available to all vessels to communicate with ships and coast stations operated by state or local governments. Messages are restricted to regulation and control, or rendering assistance. Use of low power (1 watt) setting is required by international treaty.
18A	156.900	156.900	COMMERCIAL (Intership and ship-to-coast). Same as channel 7A.
19A	156.950	156.590	COMMERCIAL (Intership and ship-to-coast). Same as channel 7A.
20	157.000	161.600	PORT OPERATIONS (Ship-to-coast). Available to all vessels. This is a traffic advisory channel for use by agencies directing the movement of vessels in or near ports, locks, or waterways. Messages are restricted to the operational handling, movement and safety of ships and, in emergency, to the safety of persons.
21A	157.050	157.050	U.S. GOVERNMENT ONLY.

continued

MARINE VHF CHANNEL ALLOCATIONS

Channel Number	FREQUENCIES (MHz)		CHANNEL USAGE
	Ship Transmit	Ship Receive	Intended Use
22A	157.100	157.100	COAST GUARD LIAISON. This channel is used for communications with U.S. Coast Guard ship, coast and aircraft stations after first establishing communications on channel 16. Navigational warnings and, where not available on WX channels, Marine Weather forecasts are made on this frequency. *It is strongly recommended that every VHF radiotelephone include this channel.*
23A	157.150	157.150	U.S. GOVERNMENT ONLY.
24	157.200	161.800	PUBLIC CORRESPONDENCE (Ship-to-coast). Available to all vessels to communicate with public coast stations. Channels 26 and 28 are the primary public correspondence channels and therefore become the first choice for the cruising vessel having limited channel capacity.
25	157.250	161.850	PUBLIC CORRESPONDENCE (Ship-to-coast). Same as channel 24.
26	157.300	161.900	PUBLIC CORRESPONDENCE (Ship-to-coast). Same as channel 24.
27	157.350	161.950	PUBLIC CORRESPONDENCE (Ship-to-coast). Same as channel 24.
28	157.400	162.000	PUBLIC CORRESPONDENCE (Ship-to-coast). Same as channel 24.
65A	156.275	156.275	PORT OPERATIONS (Intership and ship-to-coast). Same as channel 12.
66A	156.325	156.325	PORT OPERATIONS (Intership and ship-to-coast). Same as channel 12.
67	156.375	156.375	COMMERCIAL (Intership). Same as channel 7A except limited to intership communications.
68	156.425	156.425	NON-COMMERCIAL (Intership and ship-to-coast). A working channel for non-commercial vesssels. May be used for obtaining supplies, scheduling repairs, berthing and accommodations, etc. from yacht clubs or marinas, and intership operational communications such as piloting or arranging for rendezvous with other vessels. It should be noted that channel 68 is the most popular non-commercial channel and therefore is the first choice for vessels having limited channel capacity.
69	156.475	156.475	NON-COMMERCIAL (Intership and ship-to-coast). Same as channel 68.
70	156.525	156.525	Intership and ship-to-coast selective signaling only. No voice communications.
71	156.575	156.575	NON-COMMERCIAL (Intership and ship-to-coast). Same as channel 68.
72	156.625	156.625	NON-COMMERCIAL (Intership). Same as channel 68, except limited to intership communications.
73	156.675	156.675	PORT OPERATIONS (Intership and ship-to-coast). Same as channel 12.
74	156.725	156.725	PORT OPERATIONS (Intership and ship-to-coast). Same as channel 12.
77	156.875	156.875	COMMERCIAL (Intership). Same as channel 7A except limited to intership communications.
78A	156.925	156.925	NON-COMMERCIAL (Intership and ship-to-coast). Same as channel 68.
79A	156.975	156.975	COMMERCIAL (Intership and ship-to-coast). Same as channel 7A.
80A	157.025	157.025	COMMERCIAL (Intership and ship-to-coast). Same as channel 7A.
81A	157.075	157.075	U.S. GOVERNMENT ONLY.
82A	157.125	157.125	U.S. GOVERNMENT ONLY.
83A	157.175	157.175	U.S. GOVERNMENT ONLY.
84	157.225	161.825	PUBLIC CORRESPONDENCE (Ship-to-coast). Same as channel 24.
85	157.275	161.875	PUBLIC CORRESPONDENCE (Ship-to-coast). Same as channel 24.
86	157.325	161.925	PUBLIC CORRESPONDENCE (Ship-to-coast). Same as channel 24.
87	157.375	161.975	PUBLIC CORRESPONDENCE (Ship-to-coast). Same as channel 24.
88A	157.425	157.425	COMMERCIAL (Intership). Same as channel 7A, except limited to intership communications and between commercial fishing vessels and associated aircraft while engaged in commercial fishing.
WX1	—	162.550	WEATHER (Receive only). To receive weather broadcasts of the Department of Commerce, National Oceanic and Atmospheric Administration (NOAA).
WX2	—	162.400	WEATHER (Receive only). Same as WX1.
WX3	—	162.475	WEATHER (Receive only). Same as WX1.

NOTE: The addition of the letter "A" to the channel number indicates that the ship receive channel used in the United States is different from the one used by vessels and coast stations of other countries. Vessels equipped for U.S. operations only will experience difficulty communicating with foreign ships and coast stations on these channels.

you sit the winter out and never make a call, you are not billed. The average price of a three-minute call to a local shoreside station is generally less than $2. Some phone companies may charge a one-time registration fee or a small yearly renewal fee.

Some marine telephone operators announce traffic lists on VHF Channel 16 on the hour. Due to tremendous traffic on Channel 16 during the summer months, however, you may wish to monitor, with a second VHF set, your marine telephone channel to receive phone calls the second they are placed to your vessel from a shoreside station. You can do this with a small handheld transceiver.

Since telephone channels are separate transmit and receive frequencies, it's impossible to transmit ship-to-ship on vacant phone channels. You simply won't hear the other ship station respond because your receiver is tuned 4.6 MHz higher than your transmitter. If you wish to establish ship-to-ship communications through a telephone operator, however, because the two of you are well beyond VHF line-of-sight range, the phone company will patch you in together. Your signals will boom in loud and clear thanks to the phone operator "repeater." You could be as much as 1,000 miles apart yet still stay in contact. There is a charge for the service.

The other channels are described briefly in the accompanying table and are explained more fully in VHF owners' manuals, in piloting books, and in the *Marine Radiotelephone User's Handbook,* put together by the Radio Technical Commission for Marine Services in cooperation with the FCC. A copy can be purchased for $7.95 from the RTCM, P.O. Box 19087, Washington, D.C. 20036.

remote receivers that pick up distant calls.) The usual range to any Coast Guard station or marine operator is in excess of 50 miles from any good VHF 25-watt station. If you use a handheld VHF, your range may diminish to a maximum of 35 miles, but output power on a clear channel is not of primary importance to the ultimate range of a VHF signal.

The maximum range of ship-to-ship communications is substantially less because neither station has any appreciable height over the water. Average ship-to-ship communications are limited to 15 to 20 miles on most days. Given the heights of both antennas, specific maximum ranges are easily calculated: An antenna 10 feet off the water "sees" the horizon at 4 miles, and two such antennas will "see" each other at 8 miles. Adding the 22 percent factor for refraction of VHF radio waves over seawater gives us a maximum communications distance of 10 miles. Going to higher power levels will add no more than a half mile or so.

You might occasionally establish ship-to-ship communications between vessels up to 50 miles apart as a result of "tropospheric ducting." This most often occurs when a high-pressure system slips over your local cruising area, and there is little wind. A temperature inversion forms at an altitude of approximately 1,000 feet, trapping VHF radio signals below it. In Southern California the "smog belt" at 1,000 feet sometimes carries VHF signals up to 300 miles away, ship-to-ship. These are not "skip" signals— rather, they are signals experiencing an intense super-refraction. Communications with shoreside stations over 400 miles away may be possible on rare occasions.

Range

VHF radio waves travel over the water on a line-of-sight basis plus a mere 22 percent over-the-horizon allowance for "super-refraction"; that is, the VHF waves, being only about 2 meters long, will not bend around the horizon to any great extent. Ultimate range is determined by the height of your antenna over the water and by its gain. The Coast Guard and marine operators often use well-elevated shoreside antenna locations to offer excellent range to ship stations. (They also frequently use

Antennas

The VHF antenna is an extraordinary setup of phased vertical elements within a fiberglass shaft, or a precision-wound coil topped by or built into a fiberglass or stainless steel stinger. It is, in effect, a complete system, because there is no need to develop a ground network below the antenna itself— the ground plane is built in.

The most basic VHF antenna, rarely seen because of its odd shape, consists of an 18-inch vertical spike with four 18½-inch radials protruding at a 45-degrees-down angle from the base. The entire system is a half wavelength long with respect to marine frequencies at 156 MHz, as indicated by the formula:

$$\frac{468}{\text{frequency (MHz)}} = \text{Length in feet}$$

This half-wavelength ground plane made up of a quarter-wavelength vertical radiator and quarter-wavelength radials offers a feedpoint impedance ("feedpoint" meaning the point of entry of the coax cable into the antenna) of 50 ohms—exactly what your radio and coaxial cable would like to see. If we were to bend the radials straight down, the impedance would be 70 ohms and a mismatch would occur. Mismatched impedances cause reflected power and power losses.

All antenna performances are rated against this half-wavelength antenna system, which gives a zero decibel gain. In other words, 25 watts into the antenna will give 25 watts effective radiated power out of the antenna, and weak incoming signals will sound like weak signals at the receiver.

You won't find many spider-type ground plane antennas on recreational boats, but you will find base-loaded VHF antennas that offer good performance at masthead heights yet give only small amounts of actual gain. These 3-foot whips, usually base-loaded (only two models have the loading coil partway up the whip rather than right at its base), in either 1/2-wave or 5/8-wave configurations, are often referred to as 3 dB or 6 dB gain antennas, when their actual gain is closer to 1/2 dB. Despite their minimal gain characteristics, however, these antennas hoisted aloft provide outstanding performance compared with higher gain antennas mounted on deck. Providing a base-loaded whip is fed with low-loss coaxial cable (described later), its performance is excellent.

Base-loaded imported whips that closely resemble domestic antennas have recently flooded the U.S. market, carrying retail prices that look very attractive. Their reliability and durability is suspect, however. An evaluation performed by Metz Antennas, a U.S. antenna manufacturer, revealed that the coil section of imitation antennas featured extremely small loading coil wires, and the entire assembly was sealed in wax rather than a hard epoxy compound. When the antenna heated up from long transmissions, the coil would begin to deform, and the antenna characteristics could change enough to damage a VHF set. According to the report:

> Our Metz antenna is developed around a stainless steel shell that encompasses our DC shunt-fed coil wrapped around a form. Each of our coils is precisely tuned for proper resonance. This imported look-alike antenna features a chrome outside shell that resembles ours, and they even copied our whip post mounting assembly. Field strength tests of these imported units indicated greatly diminished radiation and receive capabilities as opposed to our domestically made halfwave antennas. We also found that these imports used PVC molds that could distort in heat on the masthead, diminishing performance.

While the 5/8-wave and 1/2-wave base-loaded whips look similar to one another and exhibit almost the same amount (0.5 dB) of gain over a dipole, their radiation characteristics are slightly different. The halfwave antenna requires no ground plane beneath it. It also features a slightly higher angle of radiation, which works out well in those areas of mountain peaks that the Coast Guard and marine operators favor for antenna locations.

The 5/8-wave base-loaded marine whip requires a metallic ground plane (such as a mast) directly below its base for proper radiation characteristics. This antenna has a slightly longer whip and its radiation is slightly lower to the horizon, favoring distant ship stations. Halfwave and 5/8-wave antennas perform almost identically when mounted properly at masthead heights.

A 3 dB gain is a twofold power increase, so a 3 dB gain antenna has the effect of doubling your transmitter power to stations miles away. The 3 dB antennas are constructed of fiberglass within which halfwave vertical elements—usually two of them—are stacked end to end to form a colinear array. This configuration takes some of that otherwise-wasted skywave energy and flattens it out closer to the water. The 3 dB gain antenna is approximately 54 inches long and may also feature a decoupling section, which prevents radio energy from traveling back down the coax cable as wasted power. Quality 3 dB gain antennas use brass rods for the radiating elements inside the fiberglass shaft. I received this report from a representative of the Shakespeare Antenna Company:

> We tore open a cheap 3 dB gain antenna, and all we found was coax cable stripped back as the radiating elements. We also found no decoupling circuit, which allowed this inexpensive antenna to radiate some of the signal back down the coax line. With any type of vibration, one could easily expect the coax assemblies to break apart on the inside.

In actual sea tests, I have found little difference between domestic 3 dB antennas and inexpensive imports. During survival tests, however, which included vibrating antennas and banging them against bridge overpasses when left upright aboard a trailer boat, I found that only domestic models survived without breaking. In other words, the less expensive antennas may work just about as well, but only when not subjected to the rigors of the environment.

A fourfold power increase on both transmit and receive is achieved with a 6 dB gain antenna. This antenna is approximately 10 feet long and contains up to four halfwave colinear elements stacked vertically on the inside. Again, almost all domestic antenna manufacturers use brass elements for many of the vertical sections, whereas the inexpensive imports use coaxial cable held together by solder. Again, in tests I helped conduct, initial performance of both antennas was almost identical, but in surviving water immersion, vibration, and repeated flexes, the domestic brass tubing model seemed to hold up better.

The granddaddy of them all is the 21-foot, 9 dB gain VHF antenna that gives us an effective eightfold power gain on incoming and outgoing signals. Using a 9 dB gain antenna is the equivalent of transmitting off a quarter-wave ground plane with 200 watts of power. On receive, the 9 dB gain antenna will pull in signals that the 6 and 3 can barely discern above the static.

Most 9 dB gain antennas separate in the center for shipping. Inside their 1½-inch diameter bodies are eight halfwave antennas stacked vertically in colinear array with matching and decoupling sections at the base. I have not found any imitations or inferior models, so one may assume that 9 dB antennas are all of top-quality construction. I have torn apart both a Shakespeare and a Hy-Gain 9 dB

A 6 dB antenna (9-foot whip) mounted on the side of a pilothouse. The objective is to locate the antenna as high up as possible.

gain antenna and found quality constructed brass tubing on the inside along with coax cable matching harnesses.

Some marine dealers pooh-pooh the 9 dB gain antenna, saying it fades in and out in heavy seas because of its extremely low angle of radiation to the horizon. While in theory this is true, in actual operation the 9 dB gain antenna will lay down a stronger signal than any 3 or 6 dB gain antenna, even in the heaviest seas or on the hardest roll. Quite simply, nothing out-talks or out-listens a 9 dB gain antenna.

When buying a VHF system, a sailboat owner should normally use a halfwave or ⅝-wave antenna on the mast. I would suggest an emergency 18-inch piece of wire in case of dismasting, to be mounted directly on the back of the VHF set. A 10-foot, 6 dB, transom-mounted antenna is a viable alternative, but in actual tests the zero dB gain masthead antenna will usually outperform the 6 dB antenna at water level.

The small-boat operator will be well served by a 10-foot, 6 dB gain antenna, which affixes easily with a single ratchet mount that allows the antenna to be laid down for trailering. This gives better range than a 3 dB gain, 54-inch, white fiberglass whip, which mounts identically.

For large yachts and commercial boats, the 9 dB gain, 21-foot whip makes sense for maximum range. This whip requires both an upper and lower support mounting bracket. If you also have loran or marine single sideband, the 21-foot whip will provide a perfect match.

Coaxial Cable

A marine VHF antenna is connected to the transceiver using coaxial cable. Some antennas have the coax cable built in, this cable being adequate for most short hookups. If a longer cable is needed, however (such as when an antenna is intended for masthead use), no cable is built in. For moderately long runs, use RG-8/X (or RG-8X) coaxial cable, which offers considerably less signal loss than the smaller RG-58AU (or RG-58A/U) cable. If your VHF antenna run is long and you want the ultimate in performance, select RG-213/U cable for minimum losses at 156 MHz. The commercial boat operator may go one step further and put in the semirigid Belden 9913 coaxial cable, which offers almost no loss at all over a 100-foot run at 156 MHz.

A coaxial cable is a shielded transmission line in which one conductor is mounted coaxially inside the other. Think of it as a rod-within-a-sleeve configuration. Due to skin effect, radio frequency current is carried on the outside surface of the inner conductor and the inside surface of the outer conductor. This keeps the energy within the cable from escaping except at the ends, where you want it. Generally, the larger (and more expensive) coaxial cable will offer radio energy less resistance with less loss than will smaller cable. This is simply because there is more cross-sectional area of conductor material to carry the current. Conductor losses are a direct function of the conductivity of the conductors, too, with most coaxial cables consisting of copper or copper and aluminum.

Coaxial cable impedance (usually 50 ohms) is a function of the ratio of the diameters of the inner and outer conductors. Most marine radio transceivers have an output of 50 ohms and are compatible with 50-to-52-ohms-impedance coax cable.

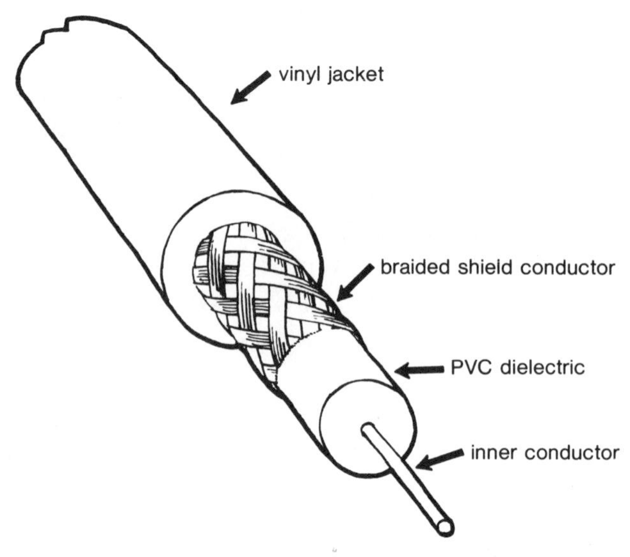

Cut-away view of a coaxial cable.

This immediately rules out using discarded cable TV coax (72 ohms), including that huge CATV coax cable that a friend snipped from a coil left outside by the local cable TV company.

Another important consideration for coax cable in a marine environment is its moistureproof capabilities. Coaxial cable jackets made of plastic alone will eventually leak moisture, but "noncontaminating" jackets made of polyvinyl chloride will resist the threats of moisture provided the cable ends have been properly terminated and sealed.

The inner dielectric (nonconducting element) can also provide a nice home for moisture, and both foam-core and air-core coaxial cables will allow moisture, through capillary action, to travel up and down the inner core and the outside copper braid. Noncontaminating coaxial cable with a solid PVC dielectric will keep moisture from getting to the center conductor and will help decrease the capillary action of moisture along the outside braid. There are some coaxial cables that not only have a solid dielectric but are also backfilled with a sticky "goo" that completely covers the outside braid within the PVC jacket to repel moisture. While this may be the ultimate coax, it's messy to work with and pretty hard to find these days.

Purchase coaxial cable only from reputable ham radio dealers—CB-type coax is out of the question. Although CB radio coaxial cable may carry the

"RG" (Radio Government) specification, the CB cable manufacturers produce a cheapened version of this coax that doesn't actually meet the original government specs. They skimp on the number of individual center conductors as well as the number of individual conductors in the braid of the copper shield. Most CB coax is quite leaky because of the huge gaps in the braid, and it distorts signals and causes impedance mismatches when the cable is flexed to any extent.

RG-8/X cable is a good size for interconnecting cables, jumpers, and short mobile runs between a handheld transceiver and a small, solid-state, linear amplifier.* Good quality RG-8/X (sometimes called miniature RG-8/U) takes the regular PL-259 antenna connector, but the slightly larger diameter UG-176/U reducer. This reducer is as common as apple pie, and you can purchase it anywhere. For a run over 20 feet, a better choice, as mentioned, is the larger coaxial cable called RG-213/U. This is exactly the same size as RG-8/U, but the designation RG-213/U signifies a type IIA noncontaminating, plasticized, synthetic-resin, protective, black jacket that will resist abrasion and will not be damaged by ultraviolet sunlight rays. You can run this cable in the bilge or submerge it in water without a chance of contamination or leakage, and it should last a minimum of 20 years. Coaxial cable without this special jacket and PVC interior may only last two years—another point against CB-type feed-lines. The larger RG-213/U is almost twice as expensive as RG-8/X (59 cents versus 29 cents per foot), but it's worth the price. At 156.8 MHz, a 100-foot run of RG-213/U causes only about 2.3 dB loss.

Transceivers

The pleasure craft operator may use fewer than 30 marine VHF channels for transmitting, but there's no real reason not to purchase a 55-channel synthe-sized VHF set that will tune in every United States channel plus a few Canadian channels at a flick of the switch. You don't pay much more for this capability. If you buy a set with 78 channels, on the other hand, you aren't getting any more U.S. channels—just a few European telephone channel capabilities. Shop accordingly. When a set boasts "101 channels," the manufacturer is using an advertising ploy; give one black mark for hyperbole.

All marine VHF sets offer 25 watts output, the maximum allowed by the FCC. In fact, thanks to imported power transistors, 25 watts is just loafing for most sets. In addition, all nationally advertised marine VHF transmitters must be accepted under Part 83 of the FCC rules and regulations. These requirements are stringent, so there is not much difference between one transmitter and another. If the set is designed for permanent installation, it will put out a clean 25 watts with good transmit audio.

Unlike the transmitter section, the receiver section of marine 25-watt VHF sets varies from extremely poor to extremely good. Manufacturers can cut costs in a VHF set by giving you an inefficient receiver. Try the set out at the electronics display of a dealer, and listen for adequate audio. Crank the volume all the way up and see whether or not it's really ear-splitting. It will need to be if you are in an open cockpit and you need to hear the set over the din of your engine at full throttle. The unit may be advertised as having 4 watts audio power, but these specs mean little. Hearing is believing. Check the channel readout—can you see the digits in the daylight? Dull light-emitting diodes (LEDs) are fast being replaced by the new bold liquid crystal displays (LCDs) that can be seen day or night.

It's pretty hard to determine receiver sensitivity as well as channel selectivity, and again, the specifications that accompany each VHF set are usually nothing more than copies of specs from their competitors. Sensitivity is the measure of the receiver's ability to hear weak signals, and you may read, for example, that a particular receiver is sensitive to a 0.35 microvolt signal at 20 dB quieting, or a 0.25 microvolt signal at 12 dB signal noise and distortion (SINAD). The greater the number of microvolts, the less the sensitivity of the receiver. Much more to the point, remember simply that sensitivity is easy to achieve, and most sets are sensitive regardless of price. Selectivity (the ability to reject

---

*The use of a linear amplifier to boost VHF output above 25 watts is illegal. A linear amplifier may be used to increase the output of a handheld unit to a maximum of 25 watts, but there are associated problems of power drain and overheating, and the cost of a handheld unit plus linear amplifier may exceed the cost of a 25-watt VHF set.

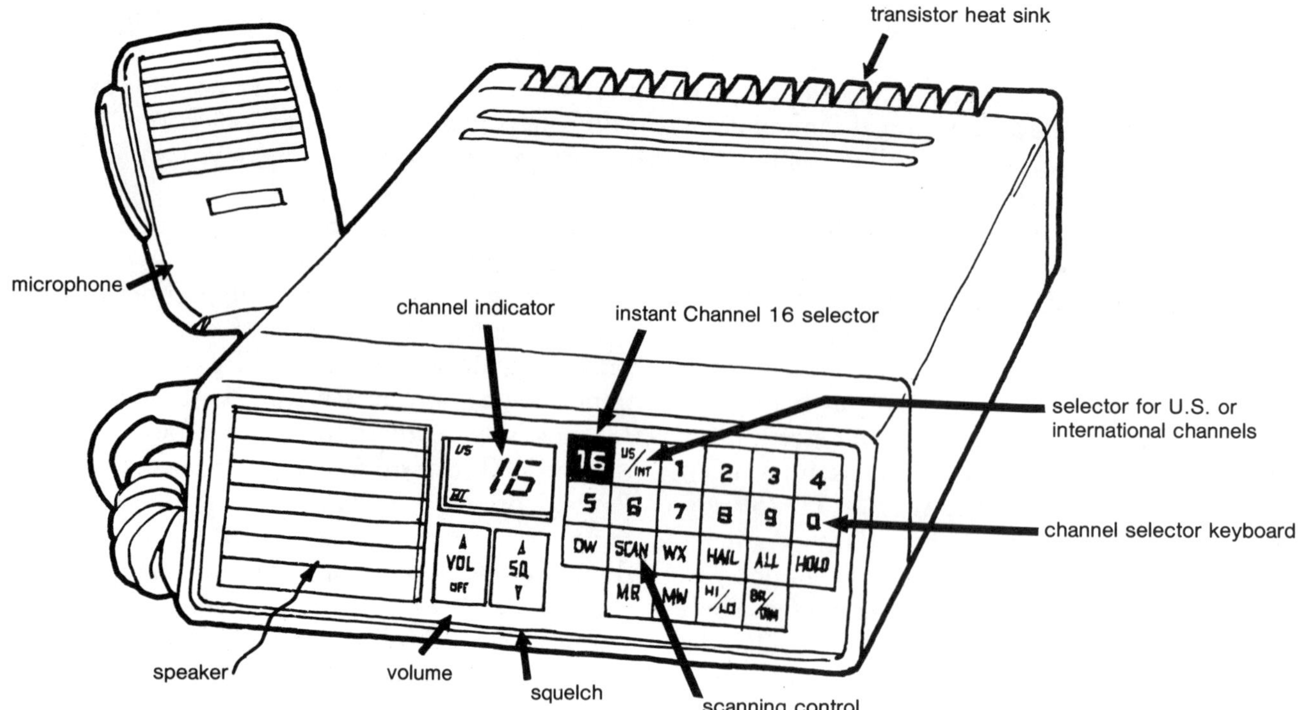

There is little standardization of display and keypad labels among keyboard VHF sets. This unit includes an automatic channel-scanning capability, connection to a loud hailer, weather frequencies, and a dimmer control for night illumination. All VHF radios must also have a switch or keypad button for low and high power output. The price range for a synthesized keyboard unit is $300 to $900 plus another $30 to $40 for the antenna. A typical size is 6 inches wide, 10 inches deep, and 3 inches high.

adjacent channel frequencies) and intermodulation rejection (the ability to reject "invading" signals formed from the mixing of out-of-band frequencies) are the critical factors in receiver quality, and unfortunately, a good way to judge these is by the transceiver's price tag. Inexpensive VHF sets may allow out-of-band signals to come through on Channel 16, so instead of hearing the Coast Guard when you're in Boston Harbor, you hear the local towing company. Quite common with cheap sets. The meaningless, copied specifications may state that both selectivity and intermodulation rejection are higher than 60 dB, but it is better to turn on the radio at the showroom and have it hooked up to the biggest VHF antenna they have on the roof. Listen to Channel 16, and if you hear strange beeps, moans, groans, and tones that come and go, chances are the receiver is experiencing intermodulation and poor out-of-band signal rejection.

The squelch circuit is another indication of quality. Try this simple test—turn the squelch off so that background noise comes through the loudspeaker at a reasonable level. Now slowly advance the squelch until the background noise disappears. Did the noise stop abruptly or did it slowly fade away as the squelch was advanced? Abrupt squelch circuits cause a DC voltage gate to switch the receiver on and off—each time with an accompanying "pop"—as signals come and go in the squelch. You can hear this "pop" even when the volume is turned all the way down by listening carefully to the loudspeaker. It is not an indication of a good, sensitive VHF set. The better receiver and squelch circuit will gradually filter the background noise as the squelch goes into its silent mode. Extremely weak signals won't pop in and out of the squelch, but will appear gracefully as smooth audio coming through your loudspeaker. Instead of a crackling noise on a weak signal, you will hear the distant voices quietly emerge.

Except in crystal sets, the transceiver is frequency selected by the synthesizer within your VHF set.

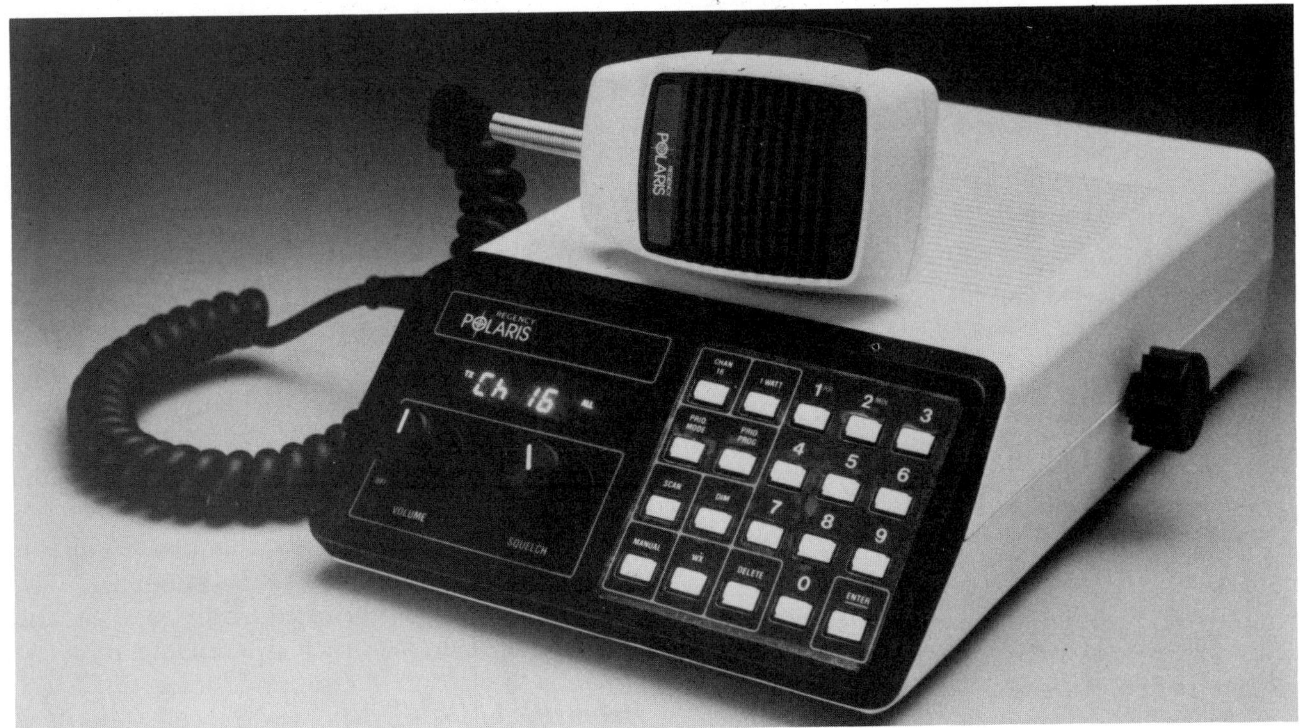

A keyboard set with typical controls.

You have a choice of keyboard channel entry or the more traditional rotating knob. The keyboard-entry VHF sets have more features, such as memorization of channels (just keyboard-enter the channels you select), channel scanning (the set monitors by turn each preselected channel, stopping on an incoming signal), channel searching (the set searches *all* the factory-installed channels), and priority channel operation (the set remains on the working channel but scans the priority channel at brief intervals, locking there if it receives a transmission). You can preprogram some units to operate at high power on certain channels and low power on others*; a few units offer a hailer-speaker accessory, which substitutes for a foghorn and can be used as an intercom.

Rotating-knob units sit on one channel unless manually adjusted. Manufacturers are producing fewer of these because of this limitation and because the mechanical contacts may become fouled or corroded in time. The keyboard units substitute miniature microswitches or pressure-sensitive membranes for mechanical contacts. Rotating-knob sets are still preferred by some mariners, however. Fishermen find the knobs easier to manipulate with gloved fingers than the small pads of a keyboard set, and a channel change can be accomplished without looking by rotating the knob through the requisite number of clicks from the previous, known channel. There are some excellent rotating-knob sets on the market.

If you plan to expose a keyboard unit to the weather, make sure the keyboard is covered by a weatherproof membrane. Some rotating-knob VHF sets are fully waterproof and can be completely hosed down without adverse effects. Behind each knob is a neoprene gasket that seals moisture out. If a radio is weatherproof or waterproof, you can be sure that the advertising literature will make a point of this. If the feature isn't mentioned in the literature, keep the set bone dry.

*Marine VHF units are legally required to offer the option of high power (25 watts) or low power output, but in truth, most mariners tend to leave their sets at high power even when talking to nearby vessels.

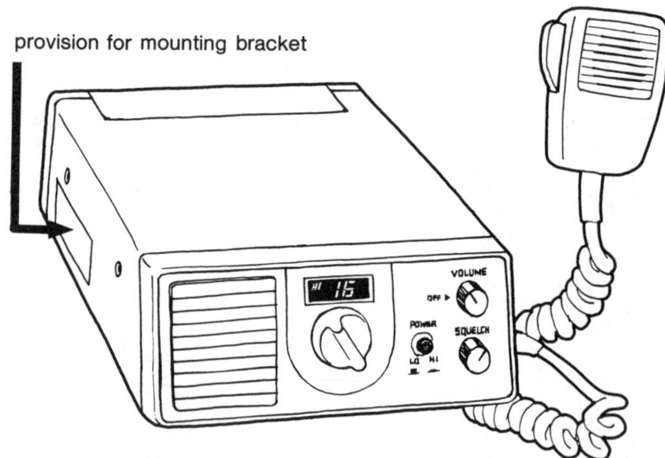

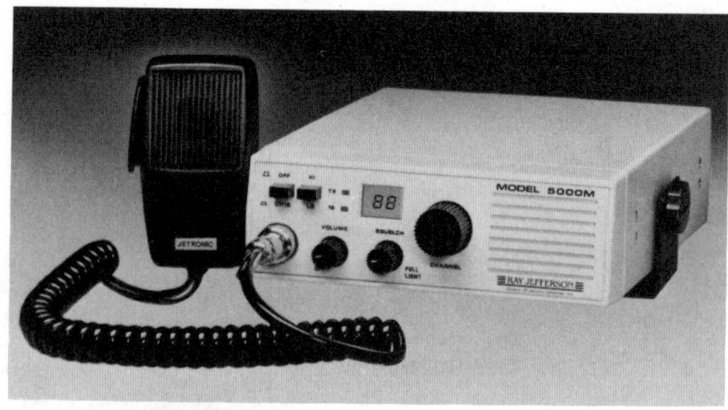

A rotating-knob VHF transceiver.

The simplest of VHF radios uses a rotary channel selector, but some such units have no more than 12 to 24 channels. That number is adequate for most boat owners unless they are cruising in international waters, where channels are used that are not authorized by the FCC for U.S. waters. Above the rotary channel selector on the unit shown here is the channel indicator, and to the right is the button for low/high power output. Farther right are the on/off/volume and squelch controls. The price range is $300 to $900 (without antenna), and a typical size is 6 inches wide, 10 inches deep, and 3 inches high.

Installation

Mount the antenna as high as possible and at least three feet away from metal spars and standing rigging except when mounted atop these. Never clamp it to stainless steel stanchions or canopy frames, since these metal objects will absorb some of the signal.

It will do no harm to run the coaxial cable with other wiring. Since the signal is kept inside the coax by the outer braid, there should be no pickup of unwanted noise. If you have excess cable, chop it off. An unnecessarily long coil of coax behind the set will only rob you of power output and weaken signal receive capabilities. If you have never soldered a coaxial cable connector (PL-259, with or without the UG-255 adapter or reducer), let someone qualified do it for you. The coaxial cable connector improperly soldered, whether at the back of the radio or the base of the antenna, is a very common failure in most new installations. As an alternative, use a crimped, no-solder connector. These are available anywhere antennas are sold, and more and more manufacturers are starting to include them with their VHF radios and loran and SatNav receivers, knowing that owner installation of this gear is on the increase. These connectors, which are simple to install, work very well but do not provide the longevity of a soldered connection. Still, they typically last about three years before the signal becomes intermittent and the connector must be replaced.

Once the cable connector is in place, you will want to test the continuity of the connection. This test cannot be done with regular ohmmeter equipment. Since most VHF antennas look to a DC meter like a direct short, it will take a specialized antenna test meter to doublecheck the operation of your antenna as well as the connection to the PL-259. While some inexpensive CB-type meters for antenna checks may work, you really need an antenna test meter calibrated for VHF frequencies. A simple way to test your antenna connection without a meter is to insert the center pin of your plug in the VHF set socket. Listen for the weak weather station signal. Now push the remaining shell of the connector assembly onto your antenna receptacle, and the weather station should increase slightly in signal strength. If it does, your connection is proper. If you slide the shell onto the connector and the weather station disappears completely, you've got a problem with your connection or with the VHF antenna itself.

In a powerboat, mount the radio where it's easy to see and operate from the helm. Mount it handy to

Right: *Proper method of installing a coaxial connector on RG-213/U cable. This cable connector is used with VHF, SSB, loran-C, and SatNav antenna cables. Proper soldering requires experience. The alternative is to use crimped, no-solder connectors, which can be installed with pliers.*

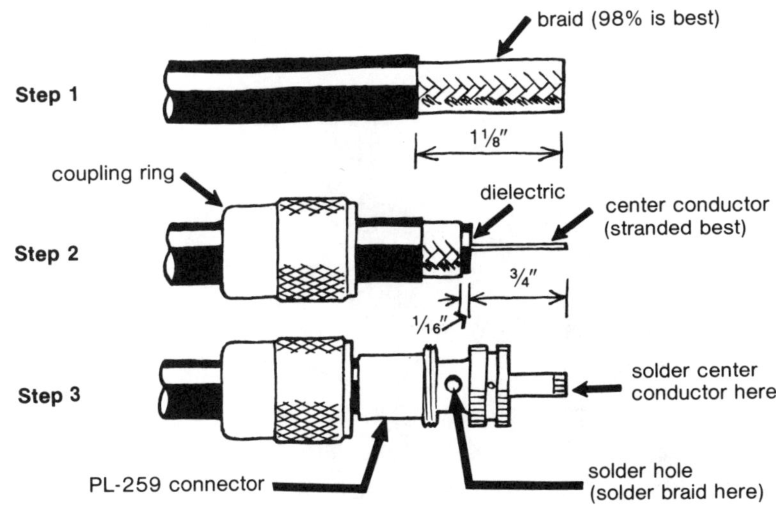

Below: *Proper method of installing a coaxial connector on RG-8/X cable. This cable is used for short to medium-length runs with VHF, SSB, loran-C, and SatNav antennas. Attempt this soldering only if you have done it before. Again, crimped connectors are the alternative.*

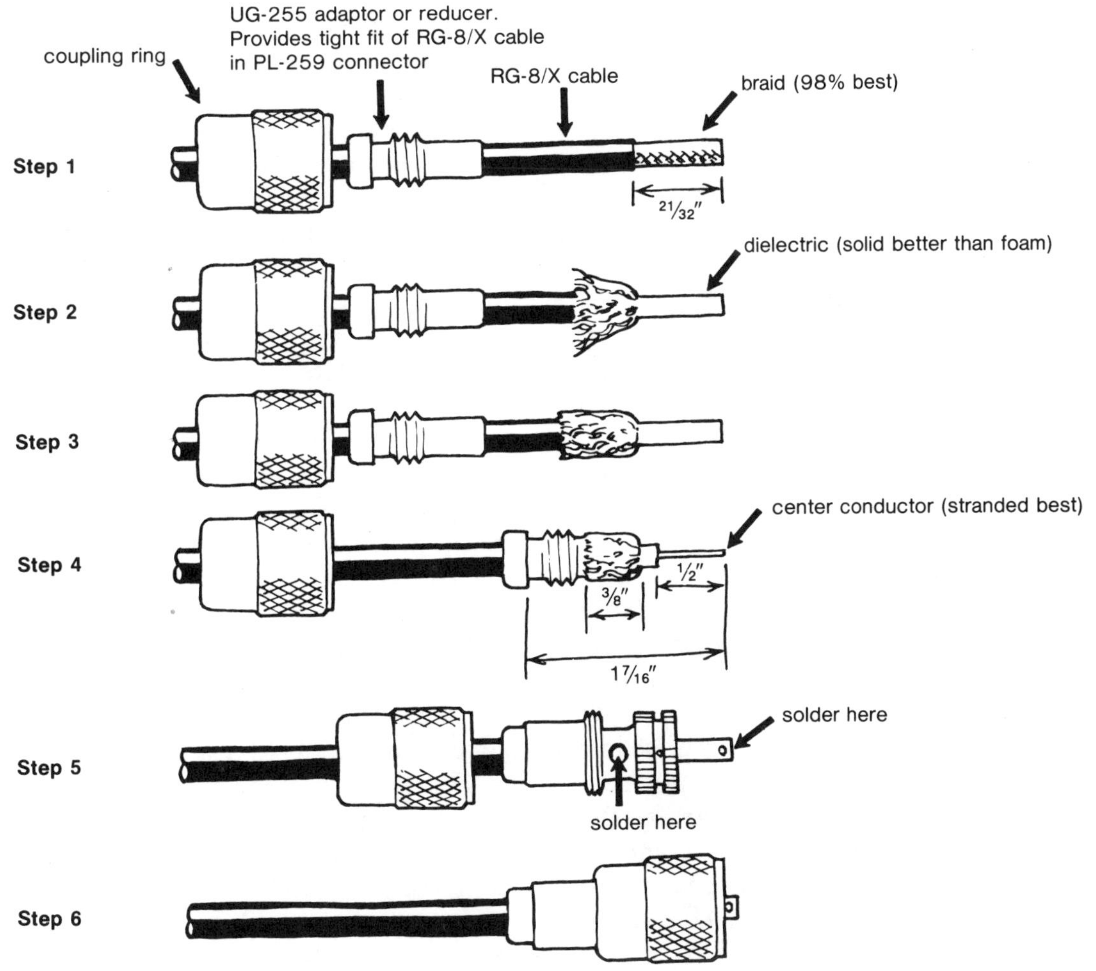

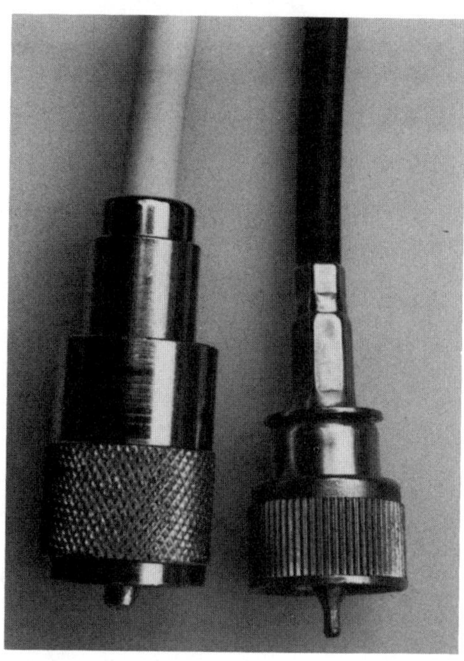

The crimp-type coaxial connector (right) *is serviceable but not as rugged and durable as a soldered connector* (left).

distant stations. Engine noise and fluorescent lights do not affect VHF radio as they do SSB and AM radios, but hot or noisy ignition circuits can produce signals that, although not apparent as background noise, will "piggyback" on weak incoming broadcast signals, confusing them. If weather channel signals transmitted 100 miles away come in loud and clear, everything is fine with your set. Don't transmit until you obtain proper FCC licensing (see below).

Handheld Units

VHF handhelds are terrific performers. A handheld is not recommended as the only marine VHF set aboard a large boat, but one of these little sets could make a great single radio installation aboard a very small boat. The handhelds are usually all-frequency synthesized. Although some manufacturers still try to peddle crystal-type VHF radios, both stationary and handheld, your better buy is a synthesized set that is just as reliable and picks up every channel.

VHF handhelds may use either keyboard-entry synthesis or thumbwheel channel selection. The thumbwheel channel selection method usually leads to a lower price tag, but the thumbwheels are very leaky and must never get wet. Keyboard entry usually leads to a more weathertight handheld that can survive a little bit of spray. No handheld will take a complete underwater dive.

Some units offer a choice of battery packs. A high-power pack may give you 5 watts, but it will wear down after an hour or two of broadcast time. Medium-power (1 to 3 watts) and low-power packs will last progressively longer. Most can be recharged.

You can hook up your handheld to an external antenna, improving its range dramatically over what is possible with the little rubber antennas the units come with. Always use an external antenna on a VHF handheld if it is the only radio aboard.

There are a plethora of accessories for any handheld, but choose only those that match your style of operation. As mentioned, you can even get an external power amplifier that takes the 3 watts output and amplifies it to 25 watts, but it's cheaper to buy a complete VHF 25-watt system in the first place.

the companionway in a sailboat. Unless it's a waterproof or weatherproof radio, keep it out of any area that might get wet. Keep any external speaker at least 3 feet from the compass. Place the microphone hanger at a convenient location, but one where the mike is not likely to be left switched on.

Run the red and black 12-volt wires supplied with your VHF set to your switch panel. Rather than tacking them onto any red and black wire you might find at your navigation station, go directly to a healthy 12-volt source that feeds voltage selector switches. Use a voltmeter to doublecheck that there is no more than a half-volt drop when you transmit on high power into a dummy load (nonradiating) test antenna network. If the dial lights dim dramatically when you transmit, the voltage input to the set is inadequate.

If you purchased your set with a remote control unit for a flying bridge, run the wires as recommended by the manufacturer. Manufacturers' instruction books are usually excellent, and you should follow their suggestions precisely.

After you have doublechecked all connections, start tuning around the channels and listening for

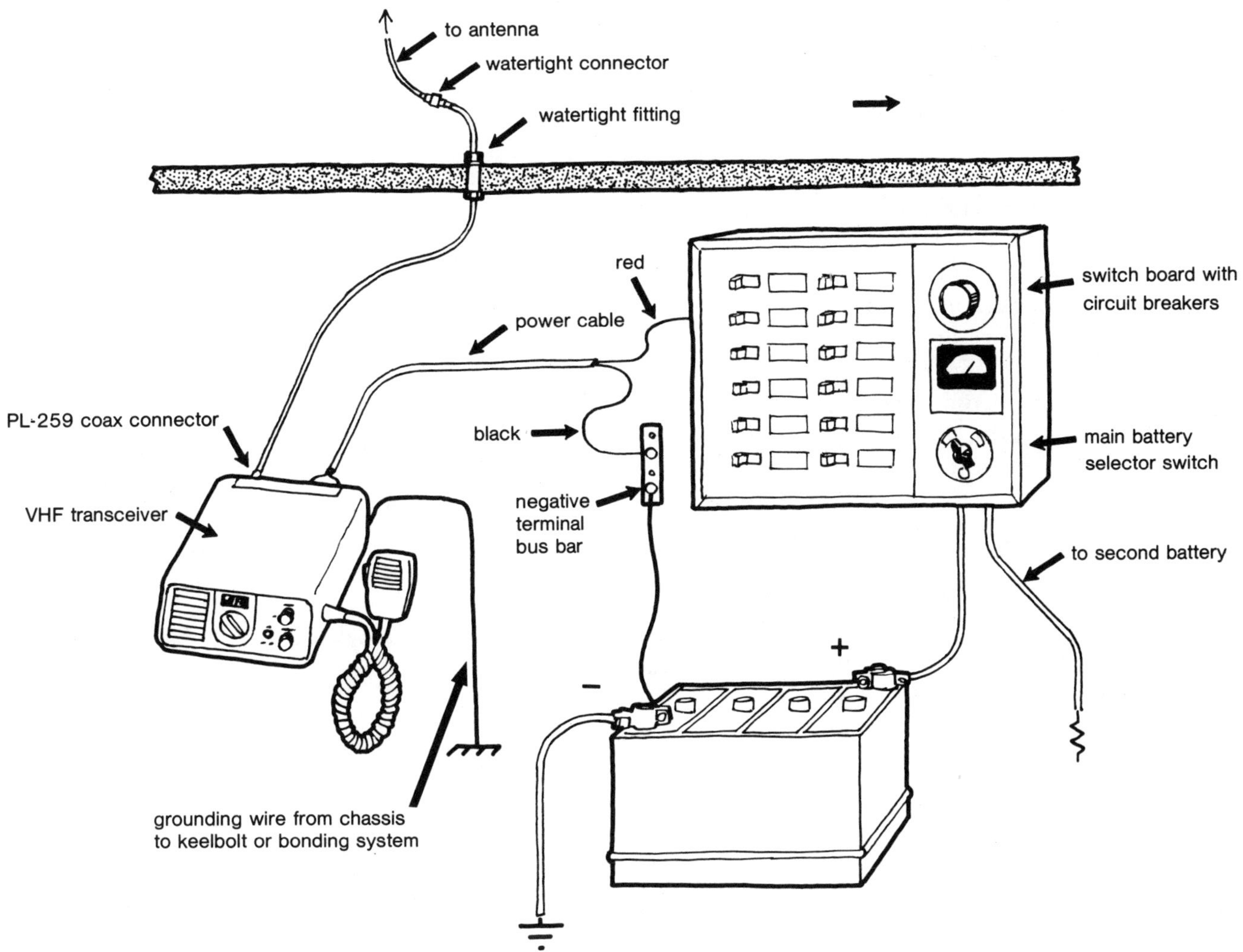

Typical hookup of a VHF radio through a switch panel. An alternative to wiring through an electrical panel is to connect the radio's power *cable directly to the battery, making sure fuses are in line with the radio.*

The licensing for VHF handhelds is identical to a regular 25-watt marine VHF set. Operating the handheld on shore is disallowed by the FCC.

Licensing

All radio stations aboard a vessel must be licensed by the FCC. You cannot transfer this license from one boat to another or from person to person. Application for a marine VHF license is on Form 506, which is normally packed with every VHF transceiver from the manufacturer. There is no fee.

The completed and signed application is sent to the Federal Communications Commission, P.O. Box 1040, Gettysburg, Pennsylvania 17325. It takes about six weeks to receive your call letters. On Form 506-A there is a special provision that allows you to use a temporary identifier as a call sign the second you put Form 506 in the mail to the FCC. Your call sign will consist of numbers and/or letters that comprise your vessel registration number or documentation number.

A personal operator's permit is no longer required for pleasureboat radio communications. The FCC recently deleted the necessity of filing FCC Form 753.

A marine radio operator's permit is, however, required for persons who operate a voice radiotelephone aboard cargo ships of 300 or more tons; aboard a vessel sailing the Great Lakes which is more than 65 feet long or carrying more than six passengers for hire; aboard a vessel carrying more than six passengers for hire, regardless of size, which is navigated in the open seas or more than 1,000 feet from shore; or from certain shore locations. If you do operate a commercial type vessel and need a marine radio operator's permit (formerly known as the Third Class operator's permit), contact the Federal Communications Commission for more information. For a radio operator test manual to prepare you for the simple multiple-choice examination, write Ameco Publishing Corporation, 220 E. Jericho Turnpike, Minneola, New York 11501 (Book No. 8-01; price $3.95).

Form 506 is also used in applying for a handheld license if you plan to use your handheld on many different vessels, but not one in particular, or if you plan to use it in an unregistered small boat.

Operating Procedures

Most instruction manuals include a radio logbook. This is a handy place to keep call sign information of stations you may communicate with, or to write down phone numbers in case you need to make a marine radiotelephone call to shore. Try to keep your log updated, and certainly note any distress messages you hear.

Most manufacturers and some marine electronics sellers include a transmitter certification card indicating that your set has been properly tuned and certified for operation. This important information must be kept with your logbook. If you find that your transmitter has not been certified by a technician, you are required by law to obtain this certification. Don't have a tech come all the way down to the boat and charge you an arm and a leg for doublechecking your transmitter frequencies, but do try to obtain a certificate from the original seller or directly from the manufacturer. The days of the FCC walking the docks and checking logs are over, but keep it legal.

Good operating procedures are necessary to keep communications flowing smoothly on the crowded

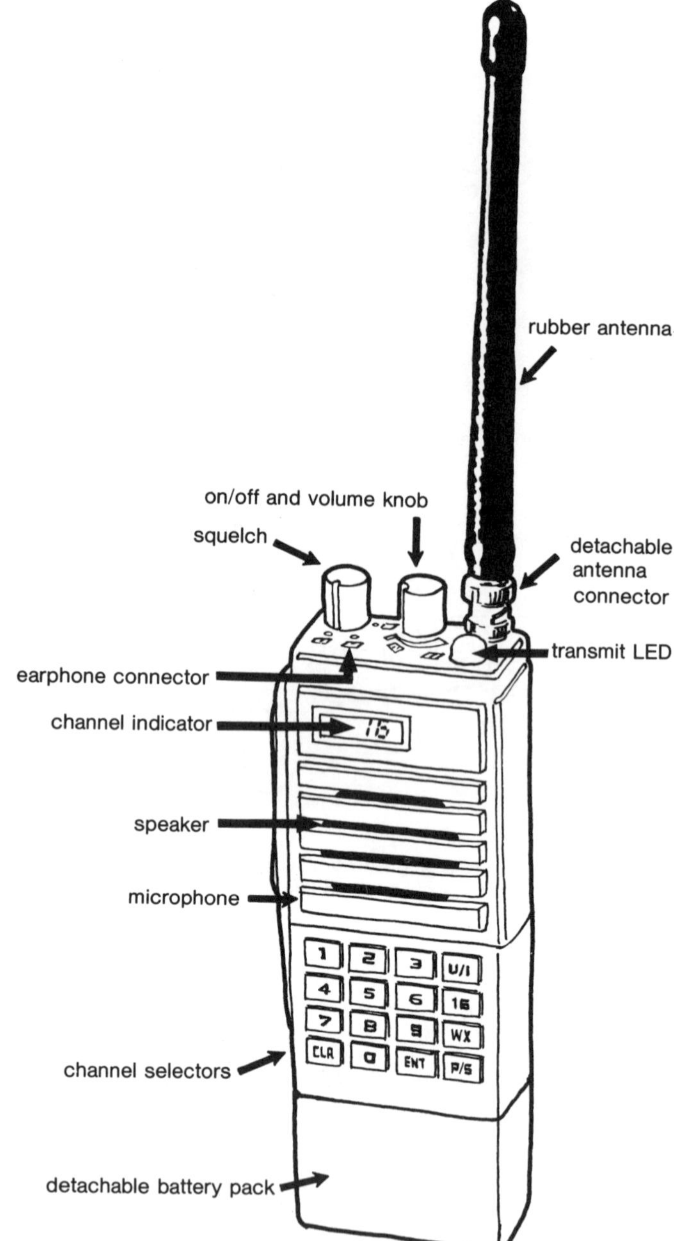

Typical handheld VHF radio with self-contained batteries. Some units are all-channel in their capabilities, while others have a limited number of channels. Range is shorter than it is with radios using a conventional antenna, although some models will take an external antenna for extended range. The price range is $300 to $600, and a typical size is 8¹/₂ inches high, 3¹/₂ inches wide, and 1¹/₂ inches thick.

marine VHF channels. Over the last few years, the VHF marine radio channels have become even more overcrowded and even more undisciplined.

A handheld VHF radio in use. Replacing this rubber-duckie antenna with a telescopic antenna would improve the unit's range.

The only time you are permitted to talk about anything other than relevant ship business is when you are tied into the marine operator. CB jargon is inappropriate and illegal. Random channel-hopping is also illegal. By law you must stick with channels that agree with your type of operation.

It is also illegal to call the Coast Guard for a radio check. They have such sophisticated monitoring equipment that any type of signal will sound loud and clear to them. Call another boat instead.

The United States Coast Guard and the Radio Technical Commission for Marine Services (in co-operation with the Federal Communications Commission) recommend standard procedures for establishing communications on marine VHF and high frequency channels. These procedures are detailed in piloting books and in VHF instruction manuals as well as government literature. Contact your local Coast Guard detachment if you can't locate the procedures elsewhere. These operating procedures should be followed.

Maintenance

Little maintenance is required to keep your marine VHF system perking along. Check all outside antenna connections now and then to make sure you have no bare wires or corroded cable plugs. Many VHF antennas bring the coax directly into their interior, so there is nothing to worry about. Base-loaded whip antennas, however, may require an outside PL-259 connector, and it's best to unscrew this connector periodically and check out what's happening inside. If the cable and connector have turned green, replace the feedline and plug.

Any sort of plug located on the exterior of your vessel should be sealed with a flexible silicone sealant. This stuff will make more of a job of removing the connector for periodic inspection, but it will seal out the weather and corrosive elements that might otherwise attack an outside antenna connection.

Check periodically to ensure that all plugs in the back of the VHF set are snug. With the radio tuned to a transmitting weather station, wiggle each wire by turn and listen for a breakup of reception. If the weather station should begin to "break up" when you wiggle a wire, chances are that wire needs to be replaced or its connection repaired. To check the mike cord, establish communications with another vessel and then wiggle the mike cord as you're talking. If your signal breaks up, you may need a complete new microphone setup. Most mike connections go bad where the cord disappears in the plastic mike or where the cord has worked loose from the chrome chassis connector.

On the water, use your VHF set in the receive mode on Channel 16 often. Not only could this very well save a life, it also helps dry up any moisture that might be inside the VHF cabinet. Transistors don't wear out, so keep your set on whenever you're underway.

Additional Buying Tips

Buy a VHF set wherever the terms are good. You can do the installation easily, following manufacturer's instructions, without help from a trained technician. FCC rules allow you to do this and specifically indicate that no FCC sign-off is required.

Wherever you buy the radio, shop for the best price, using catalogs as a good indicator. Most catalog houses pride themselves on being able to offer the most competitive prices.

Don't be misled by claims that you are receiving "a $50 tune-up value" when you buy the equipment from a company that thoroughly checks it out and peaks it for maximum transmit and receive capabilities. The factory has already squeezed every last watt out of a VHF radiotelephone and peaked it for maximum reception, and there is little that can be done by any technician to get any more performance out of the set.

You should investigate who's going to fix the radio in case it should arrive dead. Will the organization from whom you purchased it exchange it for a new one? Will they tell you to send it back to the factory for a three-week wait for repair? Will they fix it while you wait? An even exchange is the best option; some companies do this, and others don't. Buy from those who do.

A chart comparing manufacturers' specifications and prices for VHF radiotelephones appears in Appendix C.

4

Cellular Telephones

by Gordon West

Most of us, when we go cruising, are happy to get away from the phone, but there may be times when taking or making a phone call on a day off is vitally important to business. If you carry a beeper and have had it go off on the boat, only to struggle for hours to reach the office through the marine operator on your radio, you're a possible candidate for a cellular telephone.

How It Works

The cellular telephone system is a precise clone of a home or office phone, minus the wires. Direct dialing to any shore-based phone worldwide or another cellular phone in a boat or car, direct call receiving, call forwarding, conference calling, and an answer machine are available. You can tie your vessel's security system or fire or flooding alarm into the phone to alert the local harbormaster automatically. Often, the per-minute cost of a cellular phone call to or from a boat will be less than a comparable call on a marine VHF radio.

The cellular telephone service operates on 800 MHz, near Channel 88 on a television's UHF dial. A 50 MHz band embracing hundreds of channels has been allocated for vehicular, portable, and marine cellular telephone calls. Major metropolitan areas are divided into tiny radio cells that may be reassigned to another user as soon as the mobile or portable unit travels from one cell site to another. A computer-controlled main switching office (MSO) automatically hands off your conversation to a new cell as necessary, and this "hand-over" is so quick that it's barely noticeable during the conversation. All you might hear is a quick click and a slight change in voice levels. Up to 60 channels are available within each cell site, so waiting for an open channel is unusual in most places. Still, the system can get overloaded on a Los Angeles freeway during rush hour. The 800 MHz FM cellular signal is generally clear until you reach the usable end of a particular cell.

Most coastal cellular telephone cell networks have been designed to serve freeway travelers and downtown commuters with the best primary coverage available. Marine coverage is secondary, and although coverage does exist outside metropolitan areas, you will have to investigate whether your cruising waters are covered. There are literally scores of local cellular telephone providers. Call your local telephone operator and ask for information on mobile access or cellular phone services.

The control head for a cellular telephone is similar to a home phone.

A control head (left rear) and transceiver (right rear) that may be transferred from car to boat and back. The cellular interface (foreground) allows the unit to access the less expensive conventional phone system when at dock.

Since the ultrahigh frequency signals travel line-of-sight and won't bend around objects, it's quite possible to lose cellular telephone coverage in waters that are "shadowed" by land. One would expect no coverage on the back side of an island, for example, although it is sometimes possible to stay outside the shadowed area by giving at least a five-mile berth to tall island hills. Similarly, one can expect marginal reception below bold headlands and coastal cliffs. Once you get out to sea a few miles, coverage improves.

What's Available

Several equipment options, ranging from permanent installations to little handheld sets, are available. If you presently operate or intend to operate a cellular phone in your vehicle, you might investigate the possibility of having the trunk-mounted transmitter and receiver unit available for quick removal to be placed aboard your boat. A separate control head would then be permanently installed aboard the boat, enabling you to transfer your trunk transceiver equipment (which weighs about 10 pounds) to and from the boat.

For permanent marine installations, a 12-foot white fiberglass cellular telephone antenna, which looks exactly like a marine VHF antenna, is recommended. It's placed as high as possible on the vessel's superstructure for maximum range. Manufacturers of these antennas include Celwave, Shakespeare, Hy-Gain, Antenna Specialists, and Ora.

For mariners desiring permanent installation of the entire system aboard their boats, several manufacturers offer complete systems specifically designed for the marine environment. Having looked over these sets, I can't see that they differ much from a vehicular setup, except that a white control head is used and most such phones give you the ability to plug in remote heads, which would be desirable on large boats wanting more than one phone station.

Prices for permanently installed cellular telephone sets have dropped dramatically over the last four years. Top-of-the-line equipment that might have cost $8,000 originally is now available for around $3,000. Budget cellular telephone sets that were introduced at $3,000 are now as low as

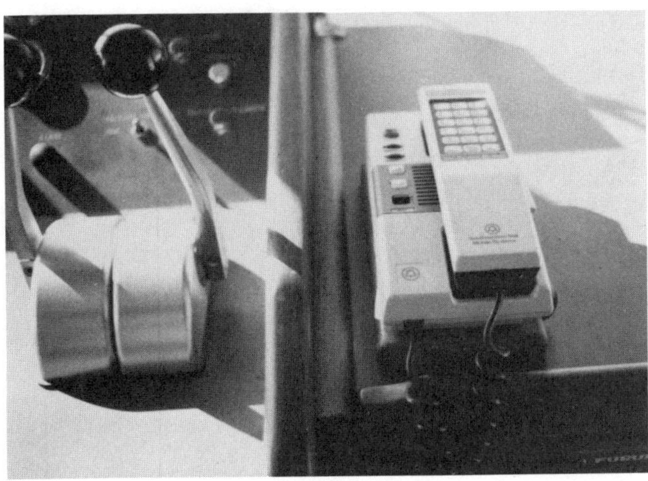

A permanent wheelhouse installation.

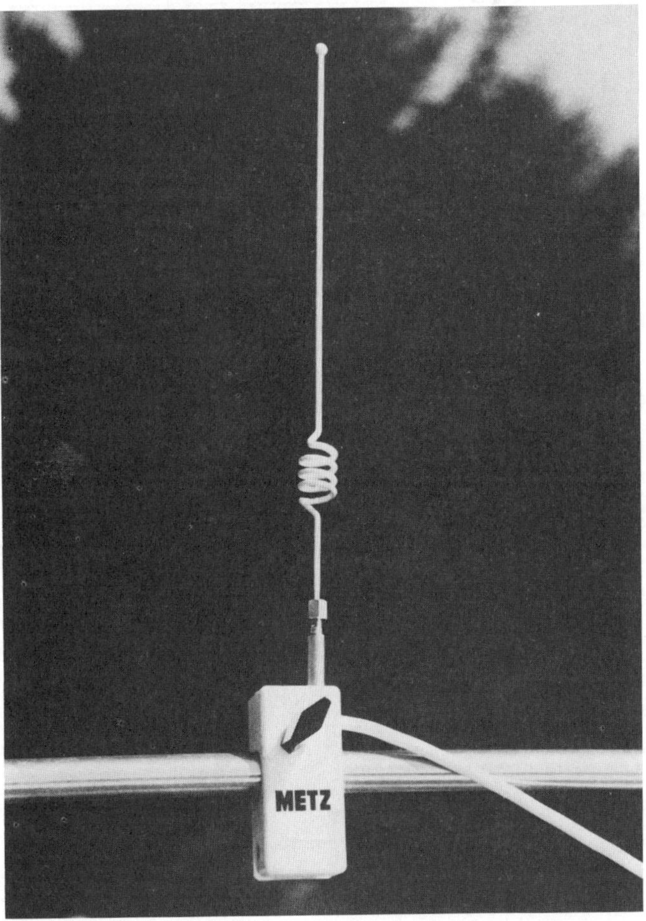

The Metz cellular antenna may be clamped onto a rail.

$1,000. The differences between high- and low-end permanently mounted sets are in their memory capabilities, the "hands-free" option, availability of add-on devices, and, in summary, the bells and whistles. All sets offer about the same power output, range, dialing capability, and transmission quality.

Nine manufacturers offer transportable, carry-aboard sets that can be tied into an external antenna system for increased range. A transportable set looks something like a large lunchpail, and it can operate up to eight hours on the hefty battery pack built into its chassis. While its little telescopic or "rubber duckie" antenna does a nice job in close, a much bigger antenna would add needed range when you're out beyond five miles or so. An antenna like the Metz antenna in the accompanying photo is frequently used in place of the 12-foot whip of the permanent installation.

A new breed of cellular phone gives you palm-of-the-hand telecommunications, but don't expect to talk for more than a few minutes between battery charges. The beauty of these handheld sets is their extremely small size—almost as small as a cordless home phone. They work well for short intervals, but you must have a way of recharging the batteries.

Call your local cellular telephone provider for more details about specific coverage areas and the best equipment recommended for your particular type of cruising.

Confidentiality

Don't expect cellular telephone calls at sea to be confidential. Anyone with a programmable hand-held, mobile, or base scanner that tunes up to 800 MHz can eavesdrop on your conversation. Marine calls are particularly susceptible to eavesdropping, because a single cell will carry the majority of the phone call. The recently enacted Electronic Communications Privacy bill imposes stiff fines on individuals caught eavesdropping on cellular calls; while it's not illegal to own a scanner that tunes in these calls, it's a crime to monitor them. Unfortunately, the new bill has done nothing more than call attention to 800 MHz phone calls and has created

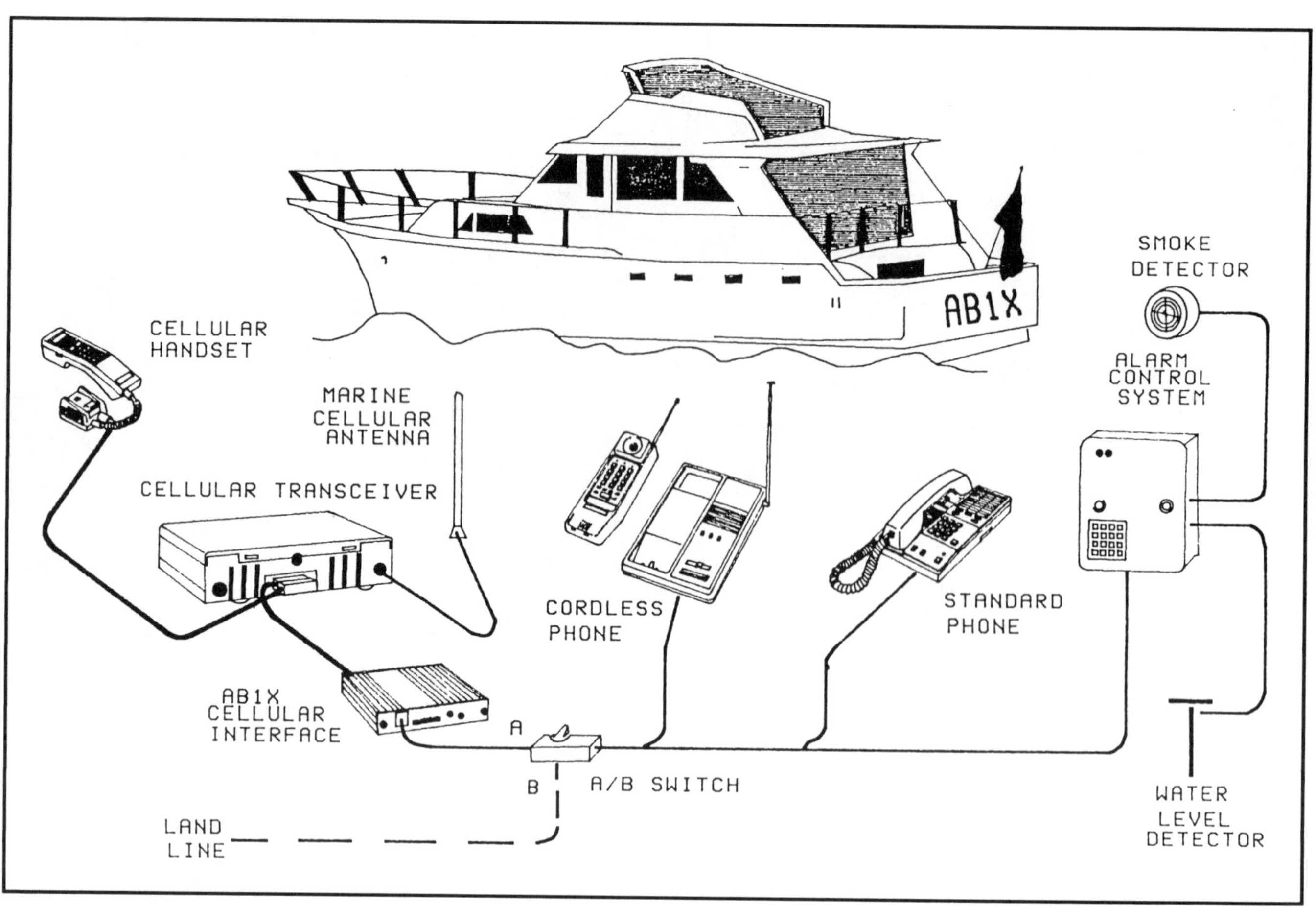

Installation of a simple two-position switch and an interface module gives a shipboard cellular telephone operator the option of receiving phone calls over his cellular telephone via the land lines when docked, or of using a standard or cordless telephone. Alarm sensors may also be interfaced with the system.

an instant market for 800 MHz scanners. One manufacturer of scanners, Radio Shack, claims they will discontinue the cellular telephone band on a new handheld model. For the foreseeable future, however, cellular phone calls are far from confidential. Try tuning your TV set on UHF Channel 88, and hear for yourself.

You can't beat the practicality of a cellular phone aboard your boat if you must stay in touch with the outside world at all times. Providing you stay inside a cellular telephone coverage area, the channels are usually wide open, with no waiting for a dial tone or retrieving a call. All you do is pick up the handset, listen for the dial tone, and start talking.

Monthly phone bills are between $100 and $150, depending on how much you talk and how many long distance phone calls you make. If you use your equipment only now and then, your bill will be below $100 per month, but unlike the marine VHF service, you are charged about $75 for monthly service whether you use the equipment or not.

A comparative chart of manufacturers' specifications and prices for cellular telephones appears in Appendix C.

5

Marine Single Sideband Radio

by Gordon West

Marine single sideband (SSB) radio enables mariners to communicate ship-to-ship or ship-to-shore over much greater distances than VHF transmissions can cover. The maximum possible communications distance is determined by the interplay of many factors, of which the time of day, the choice of marine SSB frequencies, and the quality of the shipboard installation are generally the most important. In favorable circumstances, communications around the world are possible.

If this sounds interesting, you'll also be interested to know that SSB radiotelephones are now available at substantially lower costs with more features than in the past. Marine SSB equipment is more expensive than VHF radiotelephone gear, however, and it is not a replacement for VHF. The FCC requires you to have a permanently installed VHF set aboard before you can add any other marine radio, and your SSB set should never be used when the VHF radio would suffice. To do so would only cause needless crowding on the marine SSB frequencies.

A complete marine single sideband station consists of a transceiver, an antenna, and—where channels on more than one of the seven designated marine SSB bands are to be used—an antenna coupler (also known as a tuner) between the transceiver and the antenna. Both the transceiver and the coupler must be electrically bonded to a proper ground, "proper" in this context implying elaborate and, perhaps, difficult to install. Without a good ground, the SSB station simply will not function adequately. When it works well, however, a marine SSB radio is a provider of entertainment, information, and communications for fishermen, commercial vessels, offshore sailors, and bluewater voyagers; most important, it could be a lifeline in an emergency situation.

How It Works

The term "single sideband" designates a type of radio transmission that the military and ham radio operators have been using for years to transmit messages throughout the world. In 1971 the FCC phased in single sideband transmissions for the long distance marine radio service. At the same time they introduced the expanded marine VHF service for local communications and phased out the amplitude modulated (AM) double-sideband sets. A single sideband signal concentrates your voice into a tightly compacted radiowave capable

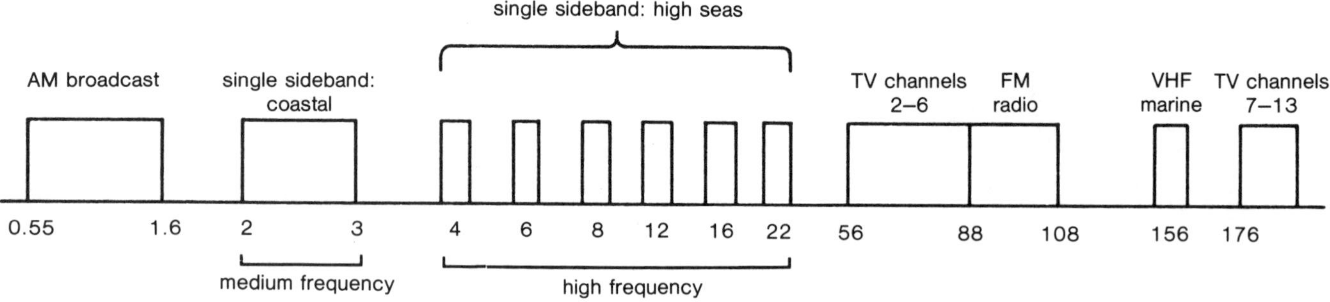

Frequency allocations at a glance.

of traveling hundreds or even thousands of miles. This very efficient radio signal is a faithful reproduction of your actual voice pattern. Unlike a commercial AM broadcast station that sends out duplicate voice waveforms plus an energy robbing "carrier," marine single sideband eliminates the unneeded mirrorlike lower sideband and the power-robbing "carrier" that has no function except to hush background noise when nothing is on the air. All the radio energy from your voice is concentrated in an upper sideband waveform that gives you worldwide talk power. Only when you speak will radio energy jump out into the airwaves, and between each word your transmitter and battery system relax. This system imposes a very low current demand on your storage battery system.

Because the transmitted signal is compacted into a very narrow band width, distant receivers are able to reject almost half the normal noise level from the air waves. FCC-required frequency tolerances keep single sideband sets precisely on frequency to minimize that "Mickey Mouse" effect on receive. Simply by adjusting a clarifier knob on your transceiver, you can produce a normal-sounding voice from a transmission by a distant ship or shore station.

A marine single sideband transceiver broadcasts on one of seven allocated frequency bands: 2 MHz, 4 MHz, 6 MHz (used mostly by stations on the Mississippi River), 8 MHz, 12 MHz, 16 MHz, and 22 MHz. The 2 MHz band is known as the medium frequency or "coastal communications" band, because frequencies in this band (between 2 and 3 MHz) are typically used for coastwise contacts just beyond VHF range. These frequencies are referred to by their actual values in kilohertz, but frequencies within the high-frequency 4 MHz through 22

MHz bands—referred to collectively as the "high seas" frequencies—are given channel designations.

When a transmission is made on any of the bands, one component of the radio signal hugs the surface of the ocean. This is called the groundwave, and it will travel approximately 100 to 200 miles. The groundwave usually does the work in single sideband communications between a boat and a nearby shore station or another boat less than 100 miles away; it is the functional component of transmissions in the 2 MHz band. Groundwave propagation is consistent day or night, depending for the most part only on a good, strong transmitted signal, though it can be interrupted by thunderstorms or other atmospheric disturbances.

The "skywave" component of the transmitted radio signal enables the long-distance communications for which single sideband radio is noted. Skywaves travel upward, bounce off the ionosphere (a multilayered envelope of ionized gas surrounding the earth), and are reflected back to earth hundreds and sometimes thousands of miles away. The height, ion density, and reflecting capabilities of the ionosphere are dependent on radiation from the sun and thus change with day and night, with the seasons, and with the 11-year solar cycle; the distance covered by skywaves of a given frequency when they reflect from the ionosphere varies accordingly. In general, lower frequencies bounce back to earth close in, while frequencies in the 12 MHz band reflect over fairly long distances, typically 3,000 miles. In the 22 MHz band you may be able to communicate from the West Coast of the United States into the Mediterranean. If the ionosphere is strong enough, the skywave will often bounce a second time, doubling the maximum possible communications distance. You can easily

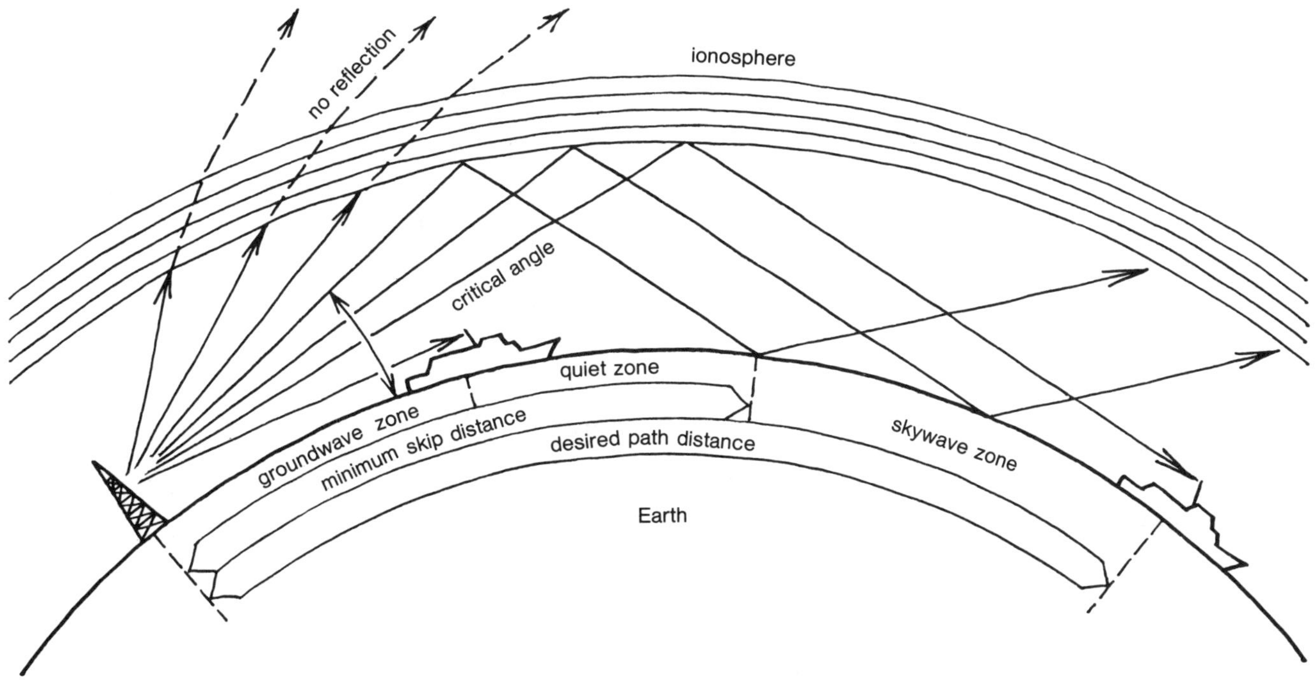

Single sideband skywave and groundwave paths.

talk all the way around the world on 22 MHz at such times.

During daylight hours the ionosphere rises and its ion density increases; the range available on the higher frequencies increases accordingly. The skywave components of lower-frequency transmissions are absorbed by the ionosphere, however, while their groundwave components are inhibited by the higher daytime noise level near the earth's surface. Maximum ranges of 2 MHz, 4 MHz, and 8 MHz transmissions tend to decrease during the day.

During the night, the ionosphere gradually lowers and its ion density decreases. Skywaves in the 16 and 22 MHz bands may not be reflected at all after the evening hours. Nonetheless, you may still be able to enjoy a communications range of several thousand miles using one of the lower frequency bands.

Skywaves are unaffected by local weather conditions.

After a few weeks of playing around with a single sideband radiotelephone, a user will begin to get a feel for the expected range on any one particular band of frequencies. As a rough guide, here is what to expect for ship-to-ship communications in the

Frequency Band	Daytime Range	Nighttime Range
2 MHz	Skywaves absorbed	1,000 miles
4 MHz	Skywaves absorbed	1,500 miles
6 MHz	500 miles	2,000 miles
8 MHz	700 miles	3,000 miles
12 MHz	1,500 miles	Worldwide in the direction of the sun.
16 MHz	3,000 miles	Worldwide in the direction of the sun until 8 P.M. local time, becoming inoperative after that.
22 MHz	Worldwide	Little skywave reflection after sunset.

groundwave range: 2 MHz—150 miles (perhaps more at night); 4 MHz—100 miles; 6 MHz—75 miles; 8 MHz—70 miles; 12 MHz—50 miles, and 16 MHz—50 miles. For comparison, a VHF radio (156 MHz) might give you about 40 miles ship-to-ship. The following table will give you an idea of what to expect from the skywave component of your SSB transmission.

To increase your maximum communication range, you should go to a higher frequency. If your signal is literally skipping over the desired station, however, you should switch to a lower frequency. In general, the greater the maximum communication range afforded by a particular band at a particular time of day, the broader the intervening skip zone.

SSB Services and Broadcasts

For any mariner outside VHF range, the marine single sideband service can be a lifeline in an emergency. The United States Coast Guard and other distress agencies throughout the world monitor 2,182 kHz (in the 2 MHz band) as the international distress frequency, allowing you to contact shoreside and marine rescue agencies immediately when outside VHF Channel 16 range. There are literally thousands of stations guarding this channel for a distress call 24 hours a day. The U.S. Coast Guard includes five additional working channels—424, 601, 816, 1205, and 1625—in their Automated Mutual-Assistance Vessel Reserve program. Imagine using your marine SSB set to place a call for help when you're thousands of miles away from any shore station. Through their AMVER system, the Coast Guard can pinpoint the position of commercial and military vessels passing through your area and signal them to change course and steam to your location to render assistance.

A marine SSB set gives access to other services besides emergency assistance. Shoreside commercial telephone stations stand by on dozens of frequencies, ready to place a mariner's phone call. These shore-based phone companies operate elaborate transmitting and receiving antenna systems to enable clear communications, and they transmit "traffic lists" to ships at sea that have telephone calls waiting from shoreside parties. They also broadcast weather reports, storm warnings, and other safety-related notices to mariners, and with their massive antenna systems they can patch a mariner into rescue coordination centers, hospitals, and emergency-at-sea medical systems, without charge, should an emergency arise.

Incidentally, shoreside marine telephone stations also make it easy to predict the best band to use for rock-solid communications. Every four hours when they read the traffic lists and give the ocean weather conditions, they transmit simultaneously on each one of their authorized bands. By switching bands while they are transmitting, you can determine which band offers the best reception. Where you hear them best is where they will hear you best.

Private shore stations share the ship-to-ship channels, meaning that a mariner can communicate directly, at no charge, with, say, a marine supply company. Other "limited coast stations" might include your own marine business, yacht clubs, marine salvage companies, and private air ambulance companies—in short, any marine business that needs to communicate regularly with distant ship stations. You can even establish a base station at your office to stay in touch regarding marine matters when you're far out at sea.

Depending on the frequency capability of your SSB transceiver, you may be able to receive other services that share frequencies adjacent to the marine band channels. For example, it is possible to tune into worldwide international broadcast stations and find out the latest news here and abroad. You can eavesdrop on the military, State Department, and foreign embassy communications that fill the worldwide high-frequency spectrum. You can tie a weather facsimile receiver into a marine SSB set and receive crystal-clear weather charts of your cruising area. You can tune in amateur radio frequencies and hear local weather reports through one of the many maritime amateur radio "nets." (If you hold a valid ham license, general class or above, you can also use a marine single sideband transceiver with broad frequency capability to transmit on any amateur radio frequency.) And you can pick up the international time signals wherever you cruise, using one of many different frequencies. Tick, tick, tick, at the sound of the tone, it is exactly. . . . You may also be able to receive transmissions originating from U.S. Air Force flights, the Strategic Air Command, Air Force 1 (the president's plane), Civil Air Patrol, United States Intelligence Agency, Antarctic stations, Interpol, U.S. weather ships, the Hurricane Research Center, and Volmet-Aviation weather broadcasts. A single sideband marine transceiver with all-frequency receive capabilities will literally tune in the world.

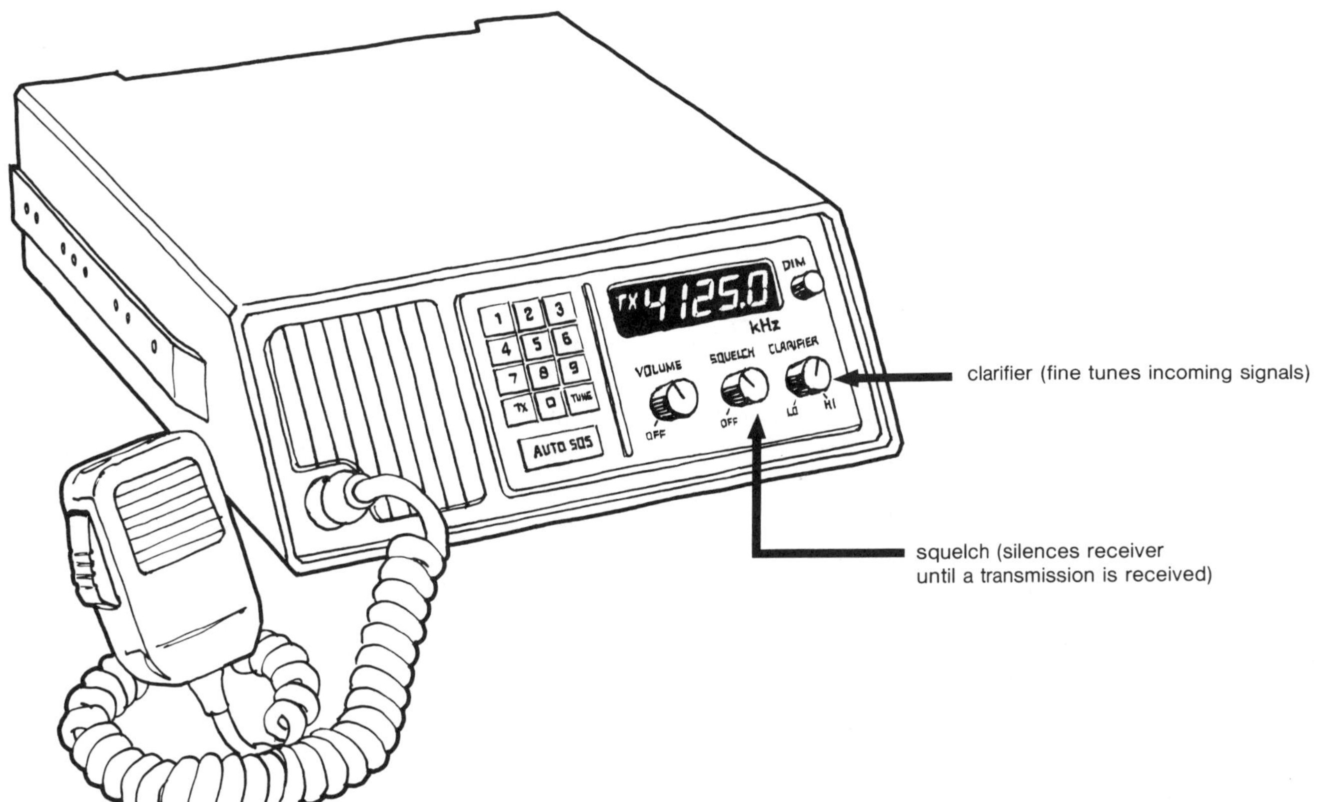

A typical synthesized SSB transceiver has controls very similar to a synthesized VHF radio. The keyboard is used for frequency selection, and to the right of it is a digital display of the chosen frequency with a dimmer control to adjust the brilliance. The price starts at $1,500 without antenna, and a typical transceiver size is 12 inches wide, 10 inches deep, and 5 inches high. If you decide on a 23-foot whip antenna, expect to pay $100 to $150.

SSB Equipment

Transceivers

In sharp contrast with most other marine electronics, the Americans have dominated the manufacturing and marketing of single sideband equipment. While a handful of single sideband manufacturers have created some excellent marine single sideband equipment, big leaps in technology and quantum jumps in price reductions have been few and far between. The switch from crystal-type to synthesized, no-crystal single sideband sets has taken place very slowly among American manufacturers, and the prices of both types have been high. Many sideband stations, complete with tuner, have been selling for well over $5,000.

In the past three years, however, the Japanese have entered the manufacturing picture, and in just one year they completely revolutionized the thinking on marine single sideband among manufacturers as well as purchasers. The very first Japanese-manufactured single sideband set, type-accepted by the Federal Communications Commission, featured 48 programmable channels, 125 watts of peak power output, capabilities for any marine channel between 2 MHz and 22 MHz, and a price tag under $1,500 not including an automatic antenna tuner and some sort of a whip or long wire antenna system. With the latter items included, the complete single sideband station still sells at half the price American manufacturers had been charging for sets with half the features.

Within 60 days after the debut of the 48-channel, completely synthesized, $1,500 marine single sideband set, American manufacturers and their dealer system dropped the price of their keyboard-entry synthesized SSB sets by $1,000. No doubt American manufacturers took a hard look at this new Japanese SSB set that's more powerful, has more channels, is priced more competitively, and is an equal performer to what was made in the U.S.A.

The price of a synthesized set with broad frequency capability is unlikely to go much lower than $1,500, but we can expect to see more Japanese-style transceivers hitting the U.S. market. The earliest entry was followed by two other sets with similar capabilities for similar prices in early 1985. There is absolutely nothing wrong with Japan's marine electronics technology. They simply have a lower paid labor force that can turn out a ton of product at an extremely low cost to a manufacturer-importer in the States. When production overtakes distribution, prices drop, and we get the benefit of high technology equipment at low prices.

In light of recent developments, a synthesized SSB set has to be considered a better buy than a crystal set. With the former it's easy to program frequencies and channels. Instead of purchasing expensive additional crystal elements—one for transmit and one for receive, a pair for every channel you want—and paying an FCC-licensed technician to install them, you can let frequency synthesis do all the work with just one reference crystal. The crystal is built into the radio, and you have only to keyboard enter the frequencies and channels you may wish to tune. Most mariners need about five frequencies in each band, and an overall total of 30 memorized frequencies is usually plenty. The older-style crystal radios offering 11 channels simply don't offer enough frequency capability to satisfy the needs of most cruising mariners.

A synthesized set that will "memorize" more than 30 channels will allow you to preprogram not only marine frequencies, but also some of those receive-only worldwide general coverage frequencies mentioned earlier. Licensed ham operators can even program transmit-and-receive marine net frequencies. Old channels can be "erased" in favor of new ones at will.

Most single sideband transceivers offer at least 100 watts output, although there are some that put out 20 watts or less. Higher output means a stronger, more solid signal provided the shipboard installation is a good one.

Antennas

After a good groundplane, the antenna is the second most important requirement for a solid transmitted signal. For powerboats, a fiberglass whip antenna is usually the best choice and easiest to install. Two models are available—a 23-foot "trap" antenna that doesn't need a tuner, or a 23-foot white fiberglass antenna, without traps, that does need a tuner. Either works well when tied into a good ground system, but most mariners choose the 23-foot white fiberglass whip without traps, and mount an antenna tuner below the antenna base. The trap antenna has high efficiency, but it is also fairly expensive—about $1,000— and usually needs to be "matched" after the installation, a job which must be performed by a technician.

Sailboat owners frequently elect to use a backstay or a hoisted wire (No. 14 plastic-covered copper wire is good) for an antenna. Either alternative is inexpensive, avoids the wind resistance and what some would consider the unsightliness of a long whip, and frequently gives better results. Insulators must be placed in the bottom and top of the backstay, but these incur no loss of rigging strength. If a dismasting occurs, however, the antenna will be lost when it is likely to be most in demand. A spare whip antenna should be carried for such a contingency.

An automatic antenna coupler is almost always necessary to tune up an insulated backstay or hoisted wire. With a backstay antenna, the coupler is mounted aft in the lazarette; a foil grounding strip goes to one side of the tuner, and the "hot wire" goes to the backstay on the other side. If you choose the microprocessor-controlled, relay-switching tuner, you will also need a tiny current draw from a 12-volt source to work the relays on the tuner.

There is no magic length of random wire for any particular frequency. The tuner will electrically lengthen or shorten a backstay or hoisted wire. Keeping the antenna in the clear is a big consideration, however. If your backstay or hoisted wire antenna is surrounded by a lot of other stays, you are likely to lose some energy into your rigging. By the same token, when a powerboat whip is mounted too close to a grounded tuna tower, valuable energy is lost.

Antenna Tuners

Unless a pretuned, prematched, 23-foot-whip trap antenna is used, any marine SSB set will need a separate antenna tuner to match the impedance of a whip or backstay antenna system to the set for frequencies between 8 and 22 MHz. (For the 2 and 4 MHz frequencies, most single sideband transceivers have built-in circuitry to couple to a 23-foot fiberglass whip or to a backstay.) The usual antenna tuner or coupler is a solid-state instrument that is weatherproof and mounts out of the way beneath the antenna system, and it can range in price from $600 to $1,500. Poor results are obtained from a $600 resistive coupler, while the $1,000-to-$1,500, relay-switching, microprocessor-controlled couplers will achieve the best possible match to squeeze the last bit of power out of the system and into your antenna setup. Top-of-the-line couplers have almost unlimited frequency capabilities.

There are also manual antenna tuners, which are available for as little as $100, but these necessitate your having to fiddle with three knobs every time you change bands. It takes skill to keep these tuners matched to the transceiver, and they are usually of interest only to amateur radio operators. (They are available only in ham radio outlets.) If you do decide on one of these, any ham radio-type 50-ohm antenna tuner will work well providing the coaxial cable feedline antenna system offers a feedpoint impedance of 50 ohms. Your set will work just fine with a manual tuner.

Grounding (And Buying Tips)

Good grounding is a must for good single sideband range, the kind of range that cruise ships, supertankers, solo sailors, the Navy, and the Coast Guard depend on. Your antenna is merely the radiating portion of the entire antenna system, and it needs to see a mirror image of itself before it will send out an SSB signal. This mirror image, called a

An automatic antenna tuner hidden in a lazarette, firmly mounted on wooden pads epoxy-bonded to the fiberglass hull. The copper grounding foil makes a solid electrical connection to the tuner's ground terminal. This tuner will work with ham as well as marine single sideband radios, and will tune any length backstay or whip within a few seconds.

counterpoise, is created by using seawater as a groundplane. The resulting balanced system may be compared electrically with a dipole antenna system—one-half wavelength long on the frequency band of operation, with voltage and current loops equally distributed throughout the half wavelength system. In marine applications using a vertical antenna, this system is more precisely referred to as a Hertz antenna setup. The white fiberglass whip is tuned to an electrical one-quarter wavelength, and the ground system will make up the other one-quarter wavelength. The ground system is the

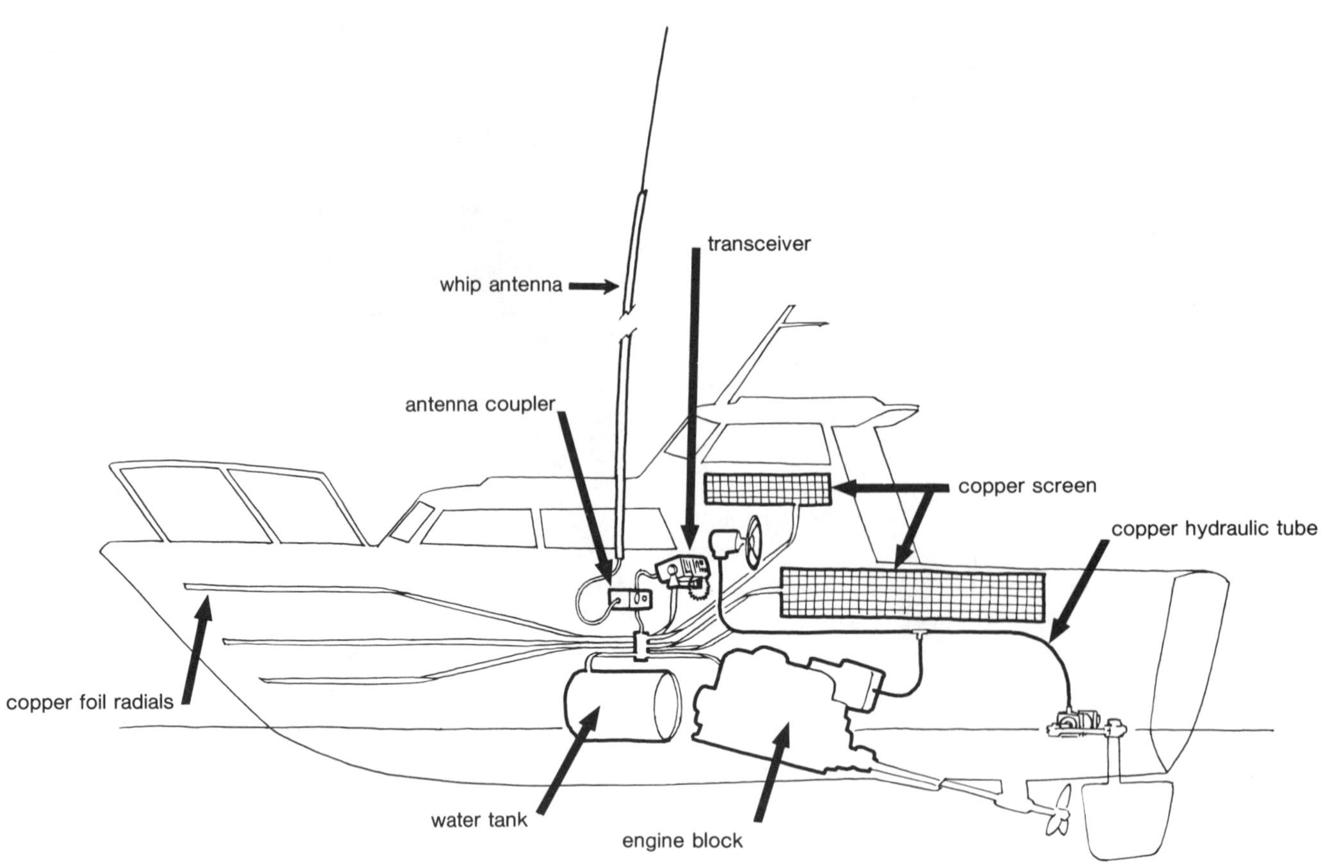

Possible components of a powerboat SSB groundplane system. Good grounding increases transmission and reception range. Copper foil works better than copper strap for radial strips and for bonding of groundplane elements.

counterpoise, and the antenna is the radiator. The effect is much like that of a diving board or the side of a pool on a diver or swimmer—the seawater gives your signal a solid surface to push off from. This is why ham radio operators would love to have seawater or a lake surface as their antenna ground, and why coastal commercial AM broadcast stations put their giant antenna systems in the mudflats. You can develop your own ground system with about a day's labor aboard your boat. The secrets of success lie in the exclusive use of copper foil and the provision of ample ground system surface area near the seawater.

Consider the consequences of an inadequate ground system. The most powerful antenna will radiate no signal if it has no counterpoise to push off from. Reception and transmission ranges will be severely reduced. In technical terms, the less ground, the higher the radiation resistance of the antenna system. This resistance will lead to power loss, and the SSB equipment will not only talk poorly but will also get quite hot in the transmitter section. An inadequate ground can actually lead to a "hot mike," wherein the operator receives a radio frequency burn each time the mike is held next to his mouth. Poor grounding will also lead to erratic movement of analog dial instruments, bizarre behavior of autopilots while the SSB set is in operation, and possible burnouts in the integrated circuits of other onboard marine electronic gear.

A boat manufacturer can add a terrific ground system when the hull is being laid up. Lightweight copper screen is one of the best ways to provide a good surface area ground, and copper screen could be laminated between the fiberglass layers. Thin sheets of copper foil could also be used. Even the conducting mesh that holds ferrocement hulls together can be used quite nicely as a ground counter-

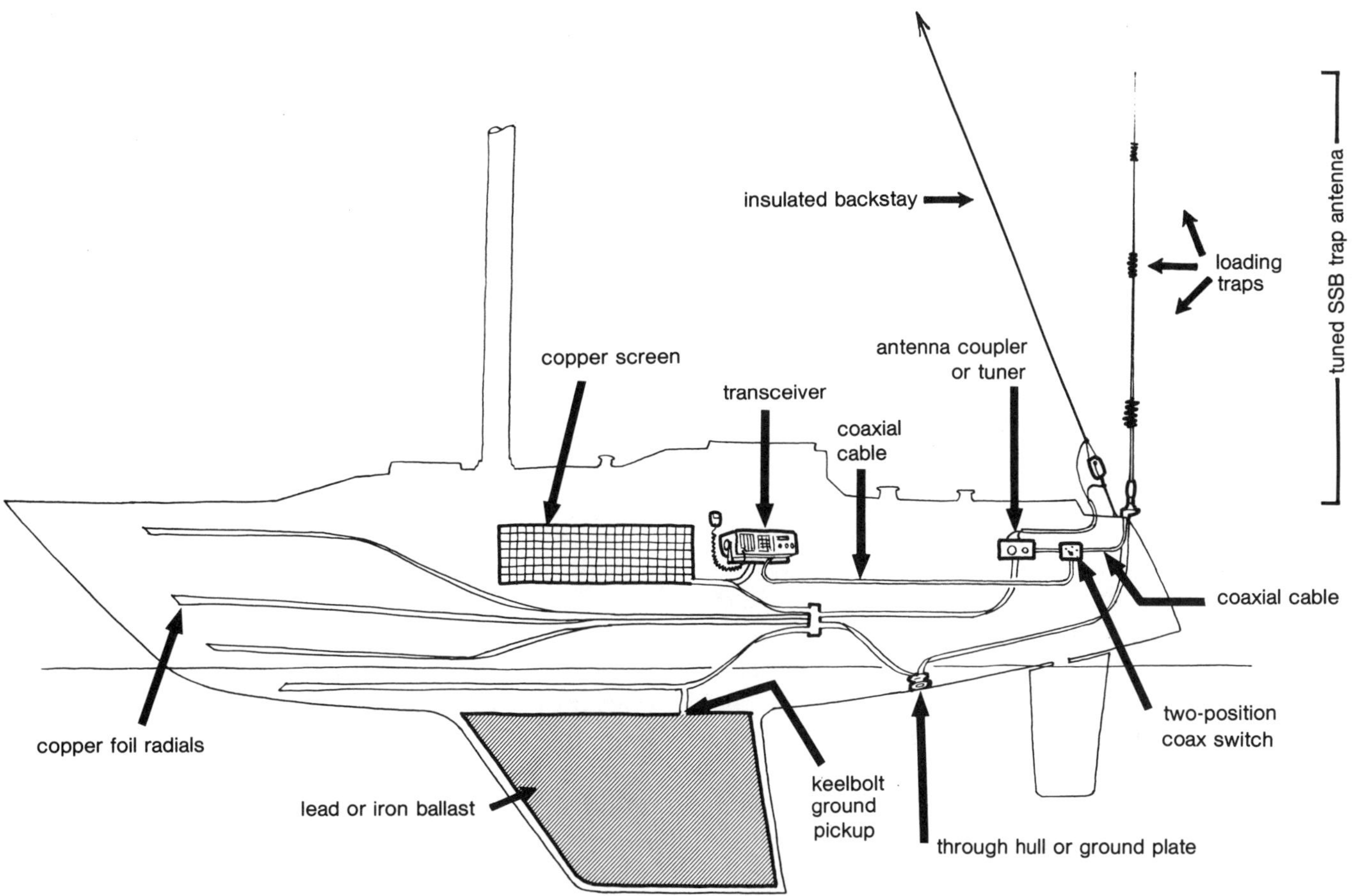

Possible sailboat groundplane system. The best and largest ground component when available is the lead or iron keel, which even if internal, provides a capacitive ground. Two alternative antenna systems may be required in racing sailboats by the rules governing offshore events. The trap antenna shown here would be installed if a dismasting were to occur.

poise system. Since grounding foil and screen is relatively expensive, however, most boatbuilders leave out the grounding systems and expect the customer to provide his own once the boat is finished. This is a shame—it's so easy to build in when the hull is under construction, and far more difficult to add after the vessel is completed.

Some technicians put ground foil and ground screen in the cabin overhead. This is true in expensive powerboats and in a few sailboats. While an overhead ground system is better than nothing, it lacks the capability of coupling with the ocean or lake as the ultimate ground system. For marine radio as well as for loran, a good ground system will comprise at least 100 square feet of metal below the waterline. A far better ground system than one in the cabin overhead would include the following, in descending order of effectiveness:

lead keel
metal water tanks
steering system hydraulic lines
external ground copper plate
100 feet of foil below the waterline
metal oil-catch pans
interconnected through hulls
foil radials
engine block

Note that the engine block is last on the list, because it does not have a large amount of flat surface area close to the water. The whole idea is surface area, and this is why a keel bolt, underwater tanks, through hulls, and anything flat and next to the water does such a good job. A good ground does not actually have to be in contact with the water, because at radio frequencies (RF), your underwater ground counterpoise sees the water in a capacitive way, which gives just as good a ground effect as if it were actually touching the water.

Another very important consideration is to use no round wires for ground. Even if you use welding wire the size of your thumb, you cannot achieve a good ground counterpoise with round wire. Round wires tend to cancel out at radio frequencies, and they look invisible as a ground counterpoise interconnect. This is why copper foil must be used between the chassis of your transceiver and your ship's ground, and between the chassis of the automatic antenna tuner and ship's ground. Good marine electronic stores that sell SSB equipment also sell 3-inch-wide, thin (exact thickness is unimportant) copper foil, as well as copper screen, for grounding. In a pinch, you can use a 1-inch-wide copper plumber's tape, but 3-inch foil is better. It takes about a day to work this foil below decks and below the waterline, picking up everything of ground potential. If you can get at your keel bolt, or tap a screw into the keel, your grounding is done. Lead encapsulated keels are the ultimate in grounds, and you need nothing further.

In a powerboat with no keel, you need to pick up as many ground potentials as possible. You can use a stainless steel hose clamp to grab each underwater metal source. If your vessel is bonded with a small green wire to each through hull, run your foil in parallel with the green wire and pick up all through hulls. You cannot rely on the green wire itself as an effective ground pickup.

As mentioned, copper foil must be run from the underwater ground system all the way up and attach directly to each piece of low- and high-frequency radio gear. While this may seem an insurmountable problem, foil handles quite nicely even in tight places. It is easily soldered to the underwater ground system and then routed up the side of the hull to the navigation station. It may be glassed into the hull, painted over, glued in, or even left exposed. The foil may be bent to a 90-degree turn, and if it *must* pass through a small hole, it may be rolled up in a not-so-tight configuration and squeezed through the orifice. Avoid a concentric-type run that tends to cancel oscillating radio frequencies. Flat is best.

There are several sticky marine compounds that will adhere the foil to the underside of a hatch or the side of a hull. Almost anything will work, and

Grounding foil will accept sharp bends as necessary, but should not be rolled like wire.

there is little danger of any substantial voltage developing on your ground foil run. Neither will your ground system substantially change your corrosion exposure to seawater. Galvanic corrosion problems occur when dissimilar metals are immersed in seawater, but the ground system is not actually immersed; its coupling is only capacitive. Electrolysis is another form of corrosion, wherein stray currents may begin to eat up underwater metals. Good wiring techniques for your 12-volt system, independent of your ground system, will eliminate electrolysis.

The 3-inch copper foil that emanates from the underwater ground system must ultimately be terminated on the instrument, but most manufacturers of marine single sideband sets (as well as loran and weather facsimile receivers) don't provide any easy way of adding ground foil to the stern end of their electronics. The best method is to run the foil up the back of the equipment and use existing sheet metal screws to make a firm connection. Where a ground post stud with nuts and a washer is provided, all the better—run the foil to the stud, double it back on itself several times for strength, punch a hole in it, and then make the connection. Don't negate all your hard work of running the foil by using a small jumper wire to connect the foil to the radio. You would be putting a "weak link" in your ground system at radio frequencies.

I usually make an accordion of the excess foil in back of the equipment so that I might remove the radio for servicing with the foil attached. If you put

Above and right: *Running the foil may take a little time and ingenuity but is not technically demanding. Three-inch-wide copper foil is recommended for SSB, ham, and loran grounding.*

bends in the right spots, the foil will resume its natural collapsed state when the equipment is put back in place. The sharp corners on the ground foil are capable of piercing the plastic protective covering on electrical wires, so make sure that red and black voltage-carrying wires are not allowed to rub against the side of the foil.

My personal preference is to ground everything at the navigation station with foil. This would include the casing of the wind and speed equipment, the autopilot control box, loran, radar, VHF, and just about everything else that lights up. The more grounding you provide for your central electronics, the fewer problems you will have with stray radio frequency noise.

The ground foil must also run to remote antenna tuners, including the tuner on a loran antenna setup as well as the single sideband tuner, which is usually several feet from the equipment. These tuners may be all the way aft, adding another dimension to your ground foil run. It's best to run the foil from the ground source directly to your tuner, rather than stringing everything out in series like Christmas tree lights. Picture one ground foil run from the keel bolt to the electronics, and a second ground foil run from the keel bolt aft to the sideband tuner and a stern-mounted loran whip. These tuner ground circuits are mandatory for reliable operation. If you try to run a sideband set with a remote tuner that is ungrounded, you stand the chance of not only burning up your equipment, but also damaging your other onboard electronics with stray RF. If it's not easy to run ground foil from your central underwater ground source aft, figure out another way to do it. It has to be done.

Again, you can pick up additional ground counterpoise surface area potential by adding substantial metals along your copper foil run. Stainless steel hose clamps make it easy to pick up bronze through-hull fittings, water tanks, copper hydraulic lines, bilge pump valves and copper lines, small underwater grounding lightning plates, and anything else that may give you some additional underwater surface area.

The engine can also constitute a small percentage of the underwater ground system. You will probably require a wiper brush if you plan to use the shaft and prop as additional ground plane counterpoise. Since the gearbox is filled with oil, and oil is not a good conductor, the wiper brush is helpful not only for grounding but also for corrosion control. Some good surface area can be obtained on big powerboat propellers.

Also recommended is to pick up all through bolts that anchor your standing rigging. By grounding your rigging you can create a zone of protection inside the rigging in case of a nearby or direct lightning hit. While lightning strikes are totally unpredictable, good grounding techniques may provide some additional safety to those inside the zone of protection. Lightning strikes on ungrounded rigging may result in discharges between stays as well as miniature lightning zaps between metals in close proximity. Good grounding will usually lead the major portion of the lightning bolt into the seawater, satisfying its quest for earth. Installers of ground systems in areas of high lightning activity should consult local lightning experts, who have developed special exit "buttons" that will pass tremendous amounts of energy into the seawater safely, without blowing out a through hull or popping holes in a lead keel.

Mariners with sailboats having poured, encapsulated lead keels, and those with metal-hulled vessels, have the easiest time obtaining a good ground counterpoise. If the keel is visible, a second nut on the exposed thread will anchor the ground foil. Seal this connection to prevent deterioration from the bilge water. It may also be recommended by local experts to tie in the aluminum mast to this keel bolt for lightning protection. The run from the mast to the keel bolt must be smooth, direct, and without sharp turns in order to pass lightning effectively into the underwater lead. Again, consult local experts.

Steel-hulled vessels provide easy attachment for the foil. Merely scrape away any protective coating from the hull, and make a low-resistance, good-surface-area contact. Again, seal this connection well.

Installing an adequate grounding system is not simple, and you may wish to have an FCC-licensed technician do the work. If you do, expect to pay your local marine electronics dealer *at least* $700 for the work (more for a more complicated installation) *unless* you buy the transceiver from that dealer and include the installation as part of the sales package. This removes the installation charge from a strict time-and-materials basis, and it's certainly a wiser course than buying the radio from a discounter and then hiring a dealer for the installation. Insist on having the installation agreement in writing, however—along with the servicing warranty terms—and insist on further work at no extra charge if the initial installation is unsatisfactory.

A compromise solution is to buy from a dealer, do the time-consuming part of the installation yourself (running copper foil and screen throughout the boat, and grounding to keel bolts, etc.), then get the dealer's technician to solder the connectors, install the transceiver, and check out and certify the installation. You'll save some money, but you'll probably also lose the guarantee of further work at no charge should the installation prove unsatisfactory.

Transceiver Installation

The very compact SSB transceiver occupies little room. Everything is contained within the chassis, which can nestle right alongside the other nav gear in a typical station. The sets are not waterproof, so keep them clear of portholes or hatches that might admit water. You can, however, build the equipment into a wheelhouse instrument panel without worry of lack of ventilation. Since SSB radios are transistorized, the slight amount of heat they develop merely dries out their insides. You may well want to keep the equipment down low for easy channel selection, making it comfortable to operate. Then, some night in a cozy harbor, you can simply flip through the worldwide frequencies to pick up some action. A heavy-duty mounting bracket is shipped with each rig.

Each transceiver is shipped with a red and black power cord. This is the 12-volt connection, and it is fused in case you should get turned around on polarity. A 150-watt marine single sideband transceiver will draw approximately 18 amps on voice peaks. Only when you talk is current consumed in these proportions—so don't worry, you're not drawing a continuous 18 amps out of your battery while the mike button is pressed down!

It's recommended to hook up your 12-volt connections directly to your ship's battery system. This allows you to stay on the air in case of a malfunction of your electrical panel, which is when you may need your set the most. If you have some hefty 12-volt wires leading from your buried battery compartment to your fuse panel, however, you may prefer to go ahead and make your connection at the panel. Clip off any large amount of excess power cable, but leave just enough coiled up behind the radio so it can be pulled from its mount for adjustments or examinations while turned on.

Route your power cable along the same track as your ground foil, taking care that the sharp foil edges don't nick the cable. Don't use the ground foil as the black side of the power cable—these are two separate "ground" systems. Routing them together creates a neat installation, and the foil partially shields the power cable from ignition interference.

Use wire lugs to attach the cable to a terminal strip on an electrical panel or, if the power cable does not already enter the console, on the back of the radio itself. Since the radio power lead is already fused, you do not necessarily need to go through an external circuit breaker. You can if you want, but that adds one more "weak link" in your power cable assembly.

If you run the power cable to your battery system, choose a battery that is less apt to fail in an emergency. A battery located above the waterline, for example, is less vulnerable to flooding.

If you need to extend the wires supplied by the factory, use No. 6 insulated two-conductor, and make certain that any splices are well-soldered or perfectly crimped and are well protected from the salt environment. Crimped (versus soldered) connectors are discussed in the chapter on VHF radios.

Eliminating Noise

Once your SSB station is installed, turn it on and start listening to the bands. There will be more atmospheric noise on the lower frequencies than the higher frequencies. With your engines and other motors all turned off, this noise is just the usual background static present on every band until a signal appears. Strong signals will usually completely mask the noise, but weak signals on 2 and 4 MHz will only quiet it by about 50 percent. The more sensitive your receiver, the more atmospheric noise you are going to pick up. This is normal; any attempt to filter out atmospheric noise would cause distant radio signals to fade away, too.

The type of noise that can be eliminated is that generated by your engine ignition system or by other onboard motors. Fluorescent lights also create noise that is usually heard on the lower frequencies. Onboard noise sources can be filtered at the spot where they are generated. There are filters for alternators and filters for fluorescent lights. You can put resistor spark plugs on your gas engine and tachometer filters on your electronic tachs. Fuel pumps and bait tanks can be silenced with appropriate filters.

Tune in a relatively weak signal on your SSB set, and then start up the engine. If the signal is still there, your noise problems are few. If the signal disappears— especially if it's a strong signal—you will need to identify noisemakers and get a filter for each. For a complete list of noise-elimination filters, see a marine electronics specialty dealer. Refer also to the chapter on noise suppression in this book.

When noises external to your boat, such as a passing skiff with an outboard, can be heard clearly on your SSB set, simply depress the noise-blanker button on the front of the radio. This will cancel out the repetitious popping sound almost completely. While it may also help dampen repetition-type noise originating on your boat, noise filters at the source are a better way to go. As in plugging leaks, you must methodically get every single one.

Going on the Air

You are required by the FCC to have a licensed marine VHF set aboard your vessel before adding marine single sideband. The same call letters are used for both, but you may need to modify your license. Dig out your marine station license and look at the authorized frequency privileges. In addition to 156-158 MHz, does it include 1,800-4,000 kHz and 4,000-23,000 kHz? If not, you will need to amend your present station license using FCC Form 506, which is discussed in the chapter on VHF radio.

Once you have mailed this form into the FCC, you may use your presently assigned FCC call letters over the air and note in your logbook that you applied for additional frequencies.

Assuming you have a synthesized transceiver, it will have been preprogrammed by the dealer or distributor that sold you the equipment. Very likely the preprogrammed channels will closely follow the recommended channelization for a particular cruising area. Reprogramming different frequencies is easy; simply refer to your owner's manual for programming instructions.

If you followed precisely the installation instructions for transceiver and antenna tuner, you probably won't need a technician to come aboard and test everything out. The instruction manual that came with your SSB lists several ways to verify full power output. Unless the manufacturer included FCC certification with your unit, however, you *do* need a technician to "sign off" in your logbook and provide the FCC certificate.

Before transmitting on any frequency, listen! In fact, spend a complete week listening to different frequencies in different bands to get a feel for the protocol of marine single sideband communications.

When listening to ship-to-ship and ship-to-private shore station calls, you generally will hear both sides of the conversation. When tuning into the ship-to-shore marine telephone station, however, you will only hear the shore station side of the conversation. This is because ship stations transmit on a separate (duplex) frequency when speaking to marine operators. When your turn to talk to a shore station telephone company arrives, the very professional marine telephone operators and their service technicians will ask you the necessary questions about where you are, who you are, and what number you want. Follow their instructions and you will have no problems communicating through the telephone service. The same holds true for emergency communications with the U.S. Coast Guard AMVER stations.

Your first call will probably be for a radio check. Since it's illegal to call the United States Coast Guard for this purpose, try the distant high seas

An antenna farm in a California harbor. The following antennas can be combined to keep your boat from looking like a porcupine: CB and VHF; VHF and loran; weather facsimile and ham; ham and marine SSB; and AM/FM radio and TV.

marine operator. Wait until they are finished with their local weather reports before giving them a call, choosing the band that sounds the strongest to you. Marine telephone companies, if they're not really busy, are more than happy to accommodate a radio check.

You can also receive radio checks from pleasure boats that you might hear on ship-to-ship frequencies. If a station sounds weak to you, they will probably say that you are weak to them. The same applies to the telephone service—if they're not coming in strong, you won't be either. Most commercial vessels will probably ignore any calls for a radio check.

Weak signals are not necessarily a result of something wrong with your installation. Sometimes ionospheric conditions simply won't favor any band. Try the next band up—and the next band down—to improve signal reports. Try a different time of day, and expect that some days you'll have better signal levels than others. Since SSB radiowaves are dependent on ionospheric conditions, it's quite normal for signal levels to change. You may also notice that signals will fade in and out on the higher frequencies—those in the 12, 16, and 22 MHz bands. Again, this is completely normal and should result in almost no loss of intelligibility during a call.

Another way to check the high-frequency operation of your equipment is to try and receive as many foreign broadcast stations as possible. These stations should come in loud and clear at the proper times but are subject to brief ionospheric fades. If you are hearing plenty of activity on these frequencies, plus strong signals from other boats and shore stations, your installation is probably fine, and you will enjoy worldwide communications with single sideband equipment.

If you do decide to have an FCC-licensed technician check out as well as certify your equipment, a local marine electronics dealership will likely be happy to send down a tech with the proper field-strength equipment to "sign off" your station (at a time-and-materials cost). If you installed the equipment yourself, the technician will check out your antenna tuner setup, doublecheck all connections to ensure that they are weatherproof, make some field-strength measurements, and exchange signal reports with distant stations. Since an electronic technician is quite familiar with the characteristics of single sideband frequencies, he'll know when your set is on the air and operating perfectly. If there is any way to squeeze a few more watts out of your system, he will do that, too.

There are approved procedures for making both normal and emergency calls on the marine SSB radio, procedures developed by the Radio Technical Commission for Marine Services in cooperation with the FCC. These are set forth in the Marine Radiotelephone Users Handbook, copies of which can be purchased for $7.95 from the RTCM, P.O. Box 19087, Washington, D.C. 20036. Your local Coast Guard station and books about piloting and seamanship are other sources of this information.

Before placing telephone calls, you will need to register your station with the appropriate marine telephone operators. If you plan to cruise extensively throughout the world, it's best to register with all four of the U.S. coast stations providing high-frequency "high seas" service. This enables you to make a phone call and have them automatically bill it. The four stations and their addresses are: Station KMI, American Telephone and Telegraph Co., P.O. Box 8, Inverness, CA 94937; Station WLO, Mobile Marine Radio, Inc., 7700 Rinla Ave., Mobile, AL 36619; Station WOM, American Telephone and Telegraph Co., 1350 N.W. 40th Ave., Fort Lauderdale, FL 33313; and Station WOO, American Telephone and Telegraph Co., P.O. Box 558, Beach Ave., Manahawkin, NJ 08050.

Each station operates on upward of 18 channels. When calling the distant marine telephone operator, remember that much of their equipment is automated, and you'll need to make at least a 30-second call in order to raise their station on the first try. Giving your vessel name and call letters once won't do it. Call the marine telephone service by their call letters over and over again, repeating your vessel name and call letters many times and giving your approximate location. This gives the telephone station technicians time to select the best antenna for your incoming signal.

TABLE OF STANDARD FREQUENCIES

COAST GUARD (USB)

Frequency Transmit	Receive	I.T.U. Channel #	Remarks
2182 kHz	2182 kHZ	None	International distress and calling frequency to all Coast Guard and rescue agencies worldwide
2670 kHz	2670 kHz	None	U.S. Coast Guard working channel
4134.3 kHz	4428.7 kHz	424	500-mile Coast Guard working channel
6200.0 kHz	6506.4 kHz	601	Gulf Coast Coast Guard working channel
8241.5 kHz	8765.4 kHz	816	Medium-range Coast Guard working channel
12342.4 kHz	13113.2 kHz	1205	Long-range 24-hour Coast Guard working channel
16534.4 kHz	17307.3 kHz	1625	Day/evening long-range Coast Guard working channel

WEATHER FAX (FSK — frequency shift keying — printer required)

EAST COAST Receive Frequency	Service	WEST COAST Receive Frequency	Service
4271 kHz	Halifax, Canada	4802.5 kHz	Hawaii
9890 kHz	Halifax, Canada	9440 kHz	Hawaii
13510 kHz	Halifax, Canada	13862.5 kHz	Hawaii
8502 kHz	Boston, Massachusetts	7770 kHz	Hawaii
12750 kHz	Boston, Massachusetts	11090 kHz	Hawaii
9389.5 kHz	Brentwood, New York	13627.5 kHz	Hawaii
11035 kHz	Brentwood, New York	8459 kHz	Alaska
4793.5 kHz	Washington, D.C.	4346 kHz	San Francisco
10185 kHz	Washington, D.C.	8682 kHz	San Francisco
12201 kHz	Washington, D.C.	12730 kHz	San Francisco
14671.5 kHz	Washington, D.C.	17151.2 kHz	San Francisco
9157.5 kHz	Mobile, Alabama	8646 kHz	San Diego
17447.5 kHz	Mobile, Alabama	17410.5 kHz	San Diego
8080 kHz	Norfolk, Virginia		
10865 kHz	Norfolk, Virginia		
16410 kHz	Norfolk, Virginia		

SHIP TO SHIP (USB)

Channel Code	Frequency	Best Time	Channel Code	Frequency	Best Time
50 miles Ship 4-A	4125.0	Night	5000 miles Ship 16-A	16587.1	Day only
50 miles Ship 4-B	4143.6		5000 miles Ship 16-B	16590.2	
50 miles Ship 4-C	4419.4		5000 miles Ship 16-C	16593.3	
150 miles Ship 6-A	6218.6	Night	10,000 miles Ship 22-A	22124.0	Day only
150 miles Ship 6-B	6221.6		10,000 miles Ship 22-B	22127.1	
150 miles Ship 6-C	6521.9		10,000 miles Ship 22-C	22130.2	
400 miles Ship 8-A	8291.1	Day/Night	10,000 miles Ship 22-D	22133.3	
400 miles Ship 8-B	8294.2		10,000 miles Ship 22-E	22136.4	
1000 miles Ship 12-A	12429.2	Day and Night			
1000 miles Ship 12-B	12432.3				
1000 miles Ship 12-C	12435.4				

TIME SIGNALS (AM)		NEWS (AM)
Frequency	Service	Frequency
2.5 MHz	WWV Ft. Collins/WWVH Hawaii	15205 kHz
5.0 MHz	WWV Ft. Collins/WWVH Hawaii	15195 kHz
10 MHz	WWV Ft. Collins/WWVH Hawaii	11925 kHz
15 MHz	WWV Ft. Collins/WWVH Hawaii	11760 kHz
20 MHz	WWV Ft. Collins	9770 kHz
5.0 MHz	JJY Tokyo	9760 kHz
8.0 MHz	JJY Tokyo	9740 kHz
10 MHz	JJY Tokyo	9700 kHz
15 MHz	JJY Tokyo	7325 kHz
3.33 MHz	CHU Ottawa, Canada	7205 kHz
7.335 MHz	CHU Ottawa, Canada	7200 kHz
14.670 MHz	CHU Ottawa, Canada	6040 kHz
5.425 MHz	Australia	
7.515 MHz	Australia	
12.005 MHz	Australia	
12.744 MHz	Calcutta	
12.763 MHz	Germany	

6

Amateur Radio for Mariners

by Gordon West

It is now easy to obtain an amateur radio license for cruising. No longer are mariners required to be electronic engineers or master Morse code operators to pass the theory and code requirements. The Federal Communications Commission has lowered licensing requirements, and obtaining an amateur radio voice-class call sign is no harder than a week with a textbook, two weeks with code cassette tapes, and finding two hams to give you that entry-level Novice class examination. New rules now allow the Novice class operator to begin using voice immediately on one worldwide band and two VHF repeater bands. In years past, the Novice was only allowed code privileges. That's all changed now, as the FCC attempts to generate more interest in ham radio and encourage mariners who have been operating on ham frequencies without a license to become legal operators.

Licensing

The FCC Notice of Proposed Rulemaking Docket 83-27 outlines a volunteer examination program in which local hams have taken over the responsibility of giving tests. Public Law 97-259 amended the Communications Act of 1934, and that allowed the FCC to make this change. No longer will exam questions be kept secret. The questions, as well as the four multiple-choice answers for each, are now public domain. These question-and-answer test guides are similar to the FAA airman's study manuals that have been around for many years. The FCC, as well as the FAA, feel that if the applicant knows all the questions and answers that could be on the examination, they surely are qualified to become a pilot or amateur radio operator. While not all amateur radio operators agree with this new philosophy, the successful passing rate of amateur radio written and code examinations has dramatically increased.

Three classifications of license are most appealing to the cruising mariner. Licenses are additive—in other words, you can't skip over a license to take a higher-grade test. All the license exams can be taken in one sitting, however, if you're good at an all-day exam session.

Novice License. This is the traditional entry-level license. The theory test will consist of 30 questions taken out of a pool of 300 published questions on elemental electronics and basic rules and regulations. The 30-question test is multiple choice and does not require the drawing of any schematics.

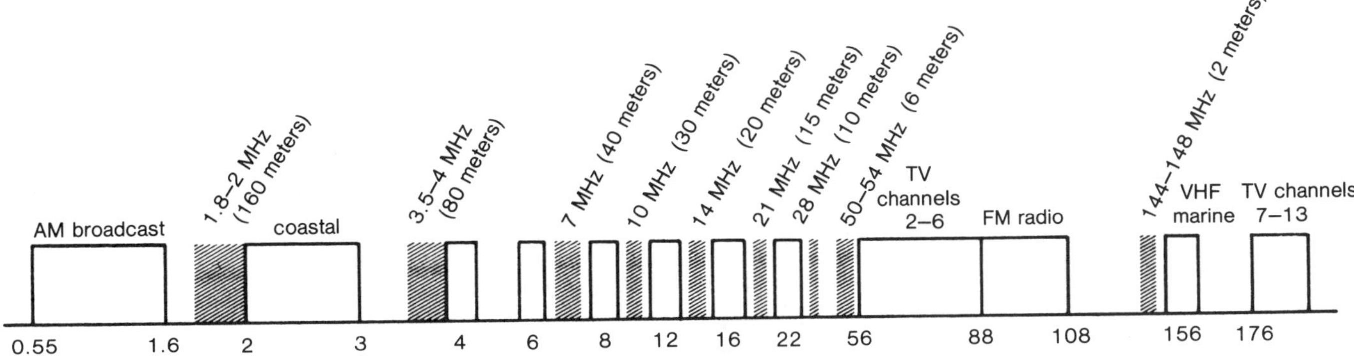

Ham radio frequency allocations.

The theory exam is easily passed with only three or four weeks of study.

A 5 words-per-minute code sending and receiving test is also required for passing the Novice test. It takes only about 30 days to learn the international Morse code at this rate.

The test is administered by any two hams who hold a General class or higher ticket and are at least 18 years of age. The volunteer hams will choose the questions from the 300-question pool in certain categories. The test may be verbal, written, or a combination of both, and the code test is simply a typical transmission of one ham to another with you copying down what was sent. You must send a few words, too, to demonstrate your ability to send code with a hand key. It takes only about four weeks for call letters to arrive.

The Novice class license holder now has voice and digital computer privileges on the amateur radio worldwide 10-meter band, on the popular 220 MHz repeater band, and on the 1250 MHz repeater band. These new privileges, in addition to the previously existing worldwide code privileges, will certainly entice more mariners into obtaining this license. The Novice class operator is allowed 200 watts of power output.

Technician Class. The Technician class license test requires no further code—just the successful passing of 25 questions (multiple-choice answers) from a pool of 250 published questions and answers. I've seen people memorize the test guidebooks over a weekend and pass the Tech test with flying colors. While I don't recommend this type of cramming, it can be done. The questions cover intermediate electronics knowledge and knowledge of the rules and regulations. If you study published question and answer test guides, the examination will be identical to the material you have reviewed. The test is taken in front of three licensed ham radio operators appointed by a volunteer exam coordinator (VEC) to administer higher-grade ham radio exams. There are volunteer examination teams all over the country, and the examinations are usually offered almost on a weekly basis. The successful passing of the examination allows immediate use of the new privileges.

Technician class operators may use voice on the 10-meter band, the very popular 2-meter repeater band, and on the 220 MHz, 450 MHz, 902 MHz, and 1250 MHz bands. On these bands are repeaters, automatic telephone patch stations, fellow mariners, and even access to the ham radio orbiting satellites. The Technician class operator may use up to 1500 watts of power output.

General Class. This is still the big one for mariners. The General class license permits voice operation on the worldwide frequencies. Worldwide ham radio nets are all centered on these frequencies, so the avid mariner should set his sights on the General class ticket.

A 25-question written test is required, as well as a 13 wpm code test. It takes about 60 days using code tapes to go from 5 to 13 wpm; you're not required to transmit—only to receive a common text as if one ham were transmitting to another. The tests are administered by three volunteer examiners appointed by a volunteer exam coordinator. These volunteers can be fellow mariners who have an Advanced class ticket or higher. While notice of the exams must be published ahead of time, they

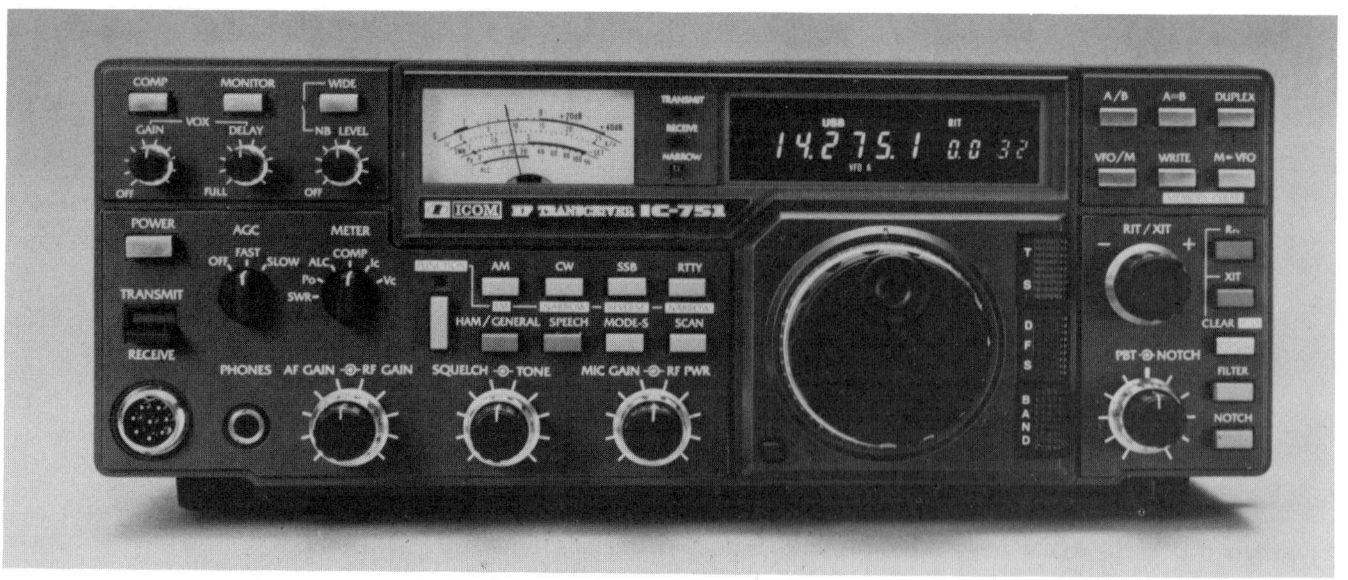

This ham radio faceplate shows an array of knobs, keys, and meters typical of ham sets but likely to bewilder many a single sideband radiotelephone operator. Implicit in this level of complexity is the assumption that the operator views his radio as more than a convenience.

can be given at yacht clubs, marinas, or even aboard cruising boats.

What happens if you should fail an examination element? Until recently, you were required to wait 30 days before retaking that element, but now the FCC has dropped the 30-day waiting period, and most volunteer examiners will let you retake any failed element the next day. Mariners may take the examination on Monday, fail it, and retake it Tuesday evening. If they fail it again, they can take it again on Wednesday, and again on Thursday if necessary. Since all questions are published, there is little chance that someone would fail it four times in a row.

A ham radio class is the best way to earn your ham radio license. Besides presenting the questions and answers on the exam, classroom study will allow the instructor to give you operating techniques, make equipment recommendations, and show you the fine arts of intercommunicating on the ham radio bands. Home study courses constitute another alternative. Courses specifically for mariners not only cover the questions and answers, but also give recommendations on marine antenna systems, grounding, and equipment best suited for the marine environment. Available from this author is a complete radio licensing course containing code and theory cassettes, theory books, a practice code oscillator setup, FCC paperwork, and a practice and real Novice class examination that you can have a fellow ham administer to you. Write to Radio School, 2414 College Drive, Costa Mesa, CA 92626.

What's Available

The "big three" in ham radio are Kenwood, Yaesu, and ICOM. While there are a half dozen other ham radio equipment manufacturers, these three offer the best value in transistorized transceivers that can easily be modified to work on marine single sideband frequencies in an emergency.

Kenwood, Yaesu, and ICOM all offer $850, compact, worldwide, high-frequency ham transceivers that tune everything between your RDF band (200 kHz) and 29.990 MHz (the top of the 10-meter ham band). These sets operate at 100 watts output single sideband, and are almost identical in their performance with the more expensive marine single sideband radios. In fact, in an emergency, as will be seen, a simple modification to your ham band set will allow it to transmit on any marine frequency to signal for help.

These $850 transceivers operate right off 12 volts. A separate power converter (for about $150) is necessary if you plan to use the unit at home on 110 volts AC. Since these 12-volt radios don't contain the big transformer power supply, their size is quite compact. They require almost no ventilation, and as long as you keep them from getting wet, they'll perform well aboard your boat. Incidentally, there is absolutely no difference between the insides of a marine single sideband set and a ham radio set when it comes to moistureproofing.

You may be asking, "Why not buy a ham radio at half the price of a marine single sideband and use one radio for two services?" For one thing, it's not legal. The FCC specifically forbids a ham set to be used on marine frequencies. Can the FCC tell if you are doing so? Absolutely not, unless they visit your boat. The frequency stability, power output, and modulation characteristics are identical. Nevertheless, the practice is strictly illegal and should be avoided. For mariners without ham radio experience, the ham set is too complicated to consider using on marine frequencies as a substitute for a marine sideband. Marine sideband sets are simple to operate—idiotproof in fact—with a minimum of knobs. Everything is preprogrammed, and you simply turn one knob for the right frequency, pick up the mike, and start talking. On the face of a ham set, there are usually no fewer than 20 knobs and buttons. Frequencies are dialed in megahertz rather than in channel numbers. Marine telephone and Coast Guard channels are all duplex (separate transmit and receive), so you would have to load two different sets of frequencies into your ham radio variable frequency oscillators in order to work these services in an emergency. Then you must remember to push the split button to have your set transmit on one frequency and receive on another. And you must, of course, be sure and select the right sideband—all marine channels are upper sideband.

The well-seasoned ham can easily dial in marine frequencies on a ham set in an emergency. Otherwise, legally, the ham set should not be used on any marine channel, although in the real world of radio the avid ham routinely misuses his set.

A tidy nav station with ham radio transceiver and other electronics.

Another choice when purchasing a ham radio setup is between a built-in or external automatic antenna tuner. One or the other is necessary for most marine installations. Transceivers with built-in automatic antenna tuners are recommended if you plan to operate mobile whip antennas (one per band) on the stern of your sailboat. Each time you change bands, you must change the mobile whip to correspond with the new frequency you plan to use. The most popular ham bands are 15 meters, 20 meters, 40 meters, and 80 meters for maritime mobile operation. You will also need specially tuned whips to cover the marine 4, 6, 8, and 12 MHz Coast Guard channels. Although these whips are "pretuned," the built-in automatic tuner will help trim out any anomalies of your ground system. (Review SSB grounding in the previous chapter.) These built-in tuners allow a perfect match to well-installed mobile whips.

Built-in automatic tuners will *not,* however, work with random wire antennas, insulated backstays, or the nonresonant powerboat 23-foot white fiberglass whip. This type of antenna system requires a "reactance-type" long wire tuner. These boxes are always separate from the transceiver, mounted way aft or directly under a powerboat whip. ICOM offers a $450 fully automatic long wire tuner, Model AH-2. With a good ground system, it will automatically tune any ham or marine frequency with any wire longer than 15 feet but shorter than 150 feet.

Unfortunately, Kenwood and Yaesu do not offer remote-mounted, fully automatic long wire an-

"Unlocking" Ham Transceivers

Maritime mobile ham radio operators are sometimes confronted with emergency situations in which their ham radio equipment becomes the only means to signal for help. The ham bands are usually full of fellow operators who can field a distress call and handle it efficiently, but in rare cases it might become necessary to place a distress call to marine radio services, such as those provided by the U.S. Coast Guard and the high-seas marine telephone operators, just outside the ham band limits.

Transmitting on frequencies outside authorized ham band privileges is strictly illegal, even though you might possess marine call letters, a CB radio license, or an aeronautical license. FCC type-accepted equipment is required for these other services, and ham radios are not FCC type-accepted. The information here is intended to assist you to take full advantage of your high-frequency single sideband set or 2-meter handheld in an emergency where life or property is at stake and you can't raise anyone on the ham frequencies.

SSB sets with digital logic are usually capable of transmitting and receiving outside the ham bands. These sets are advertised as having "general coverage" receivers. SSB sets with variable-frequency oscillator (VFO) tuning, however, don't possess general-coverage receive capabilities, nor can they transmit out of band. When it comes to 2-meter mobile sets, only a few offer expanded frequency capabilities.

Transmitting nondistress calls outside ham frequencies could result in stiff fines, possible imprisonment, and loss of your ham radio permit.

ICOM Modifications. The mobile ICOM 2-meter IC-28A and IC-28H can be expanded to 173 MHz simply by snipping diode #21 inside the top of the unit. Sensitivity at the top end is not great, but nonetheless, you have full transmit and receive capabilities. It won't go below 140 MHz into the air band, because aeronautical communications are all AM, not FM.

The ICOM IC-02AT handhelds can also be modified for extended operation, but their modification requires intricate soldering of ant-sized diodes that should only be accomplished by hams and technicians with advanced soldering skills. There are two different modifications for units before and after serial No. 34,000. Send $5.00 to Gordon West, 2414 College

Drive, Costa Mesa, CA 92626, and I'll mail you the necessary schematic prints and the programming diodes. Be sure to state the serial number of your ICOM.

On the VHF and UHF ICOM commercial handhelds, you can program frequencies on which you are authorized to transmit by unlocking the lock-out: Depress the function key, and then press numbers 1, 5, 9, 3, 5, and 7.

On the ICOM IC-735 high-frequency rigs, simply snip the teflon-covered diode leads D33 and D34 that are accessed by tearing everything apart and getting into the top side of the main PC board.

On the ICOM IC-745 HF rig, remove the covers, locate the RF board on the side of the radio, and then cut J7 pin 1, the light brown wire. On the IC-751, follow the same procedure, but locate J2 pin 1, the brown wire, and cut it.

Kenwood Modifications. The big Kenwood 940 base station is modified by first locating IC 109. Then find diode 130 and cut it for all-band transmit. If you just want MARS coverage, locate IC 111 and snip diode 135 beside it. Press "A=B" while turning the equipment on to reset the microprocessor.

On the Kenwood TS-430, locate RF unit connector #10. Simply unplugging this connector does the trick, but the Kenwood factory indicates that a safer modification would be to clip D39 near this connector and then locate IC-2 on the control unit (under the IF unit) and locate resistor R-148 near IC-2, and clip it.

On the TS-440, remove the top and bottom covers, and then remove the five small, round-head screws that secure a shiny shield plate on the control board directly in back of the front face. Then remove the two upper screws on the front panel, and loosen the bottom two screws to let the panel swivel open. Now take out the shield plate to expose the innards of the control, and cut diode 80. You can find diode 80 by looking on Page 24 of your instruction manual for the 10 Hz resolution, which requires cutting diode 66. Go ahead and whack diode 66 at the same time. Now reassemble the 440, plug in the power, and reset the microprocessor by turning the equipment on while holding the A=B button.

As for the Kenwood handhelds, the TR-2600 will go up as high as 159.995 if you cut diodes D32 and D33 on the RX unit. They can be spotted because they were originally cut, and then soldered back. After cutting, press the reset switch. The TR-2500 won't go beyond 149.99, so it's not worth the effort to try to modify it. On the 2400, no modification is available.

In the mobile TM-2500 series (2530, 2550, 2570), you can get the radio to go only as high as 150.995 MHz by cutting diodes D8 and D11 on the control unit board just to the left of IC-3. Be sure to reset the microprocessor by holding the PS key while turning the power on.

On the TS-711, you can cover MARS or CAP frequencies by cutting diode 30 on the control unit and then resetting the microprocessor.

I've been told you can get the Kenwood 3200 to go as high as 154 MHz by cutting diode 16 and adding diode 15, and then resetting the microprocessor.

In the Kenwood 830, 530, 130, 120, 820, and 520 series, no modification is available, because these sets use conventional VFO tuning. Adding crystals for fixed channel operation would be a hassle, and you should instead buy a digital-type transceiver in the first place.

Yaesu Modifications. The popular Yaesu 757 transceiver is easily modified by flipping a tiny switch that is hiding under a wire bundle inside the front of the set near the frequency selector mechanism. You will need to prowl around to find this horizontal switch, but once you see it, you can't mistake it. Just flip it in the opposite direction.

The new Yaesu microhandheld can be modified for extended transmit by snipping jumper #7, which is located in about the 10 o'clock position relative to the big CPU chip (FT-23). None of the other Yaesu handhelds will go into the 150 MHz band easily with modification.

On the new Yaesu 767 super base station, you can modify the high frequency section for all-band transmit by totally disassembling the upper and lower portions of the unit to expose its interior main printed circuit board. Then look around for a plate with holes in it for tuning slug functions and VCO caps. Peer into each and every hole with a microflashlight, and you'll probably discover that one hole is not for tuning, but rather for access to an incredibly small switch. Flip the switch, reset the microprocessor, and you are in business. No modification is available yet for those VHF and UHF modules that add to this super base station.

There are other modification techniques for different pieces of equipment. If I haven't mentioned your unit here, I either don't have the modification information, or it simply can't be modified. Please keep me up to date by sending all information and inquiries to Gordon West, 2414 College Drive, Costa Mesa, California 92626.

Remember, transmitting on unauthorized frequencies could lead to stiff fines, possible imprisonment, loss of your ham radio license, and the chance of life-threatening interference with authorized radio services. Don't play games unless you have an emergency and this is your last resort for signaling for help.

tenna tuners. You must choose the Stephens Model 1612 fully automatic tuner that works with any marine or ham SSB set. This $1,000 box does it all, but is twice as expensive as the ICOM because it will match any type of ham set.

Installation Tips

The method of installing a ham radio is identical to that for a marine single sideband set. You need a good, hefty 12 volts, a superb ground system (see the single sideband chapter), and an antenna system that is either pretuned using mobile whips or a dipole, or automatically tuned with a long wire antenna tuner if you plan to use a single, nonresonant whip (for powerboats) or an insulated backstay. Attempting to tune up your standing rigging is hazardous, although it could work in a pinch, and the same applies to trying to tune up your lifelines with a manual tuner. Although it might work, the results would be unpredictable, and there is the likelihood of a nasty burn if anyone were to touch the backstay or lifelines while you were transmitting.

The modern ham radio allows you to tune in any frequency in the shortwave spectrum with ease. You can receive weather facsimile broadcasts, 24-hour English language news, military and Coast Guard weather reports, international "spy" broadcasts, and just about anything else on the radio dial. You can also use a ham set to tune in the AM broadcast band, the Citizens Band, and the marine radio beacon service.

Amateur Radio Considerations

Ham radio is not a replacement for a marine SSB radio. A look at what ham radio can and cannot do for the mariner will clarify this frequent point of confusion and summarize this chapter:

- There are numerous maritime mobile "nets" through which ham radio operators who share similar sailing interests stay in touch. These informal nets get together at designated times of day and exchange weather

In these installations, the ham transceivers are paired with manual tuners for insulated backstay antennas.

reports, traffic lists, and other information, and shoreside ham stations are happy to help their counterparts far out at sea by "patching" a phone call to home through the land telephone lines or by ordering emergency parts. In fact, shoreside ham stations can be of great help in any emergency situation.

- Amateur radio operators engaged in local and coastwise cruising frequently carry 144-to-148 MHz VHF/FM radios (commonly known as "2-meter rigs") aboard. These sets are almost identical in size to the 25-watt marine VHF sets, and the ham VHF set can be tied into the marine VHF antenna system with a coaxial, 2-position switch.

A ham set withdrawn from its console, revealing a glimpse of the grounding foil behind. This is a good installation: The unit is comfortable to operate in its normal position; it is easy to remove for servicing; and the console protects it and the other electronic gear.

A manual antenna tuner can be used with a single-side band transceiver, but SSB operators usually choose a more expensive automatic tuner as a matter of convenience. Many hams, on the other hand, don't mind fiddling with these three dials to match the impedances of antenna and transceiver every time they change bands. This device can be tough to tune, however, so hams, too, often settle on automatic tuners instead.

The ham operator can use his 2-meter set to call shoreside stations directly or through "repeaters" and get access to local phone patches. Local telephone calls patched this way are free.

- If you're only looking for a way to avoid the marine operators on VHF or SSB frequencies and make cheap phone calls to your home or business, forget it. No business calls are allowed, and that includes the routine (as opposed to emergency) ordering of parts or supplies. Calls must be of a personal nature. Furthermore, ham radio is a self-regulating hobby, and the primary purpose of obtaining a ham license is to share the enjoyment of communicating with fellow operators worldwide. Shoreside ham radio operators aren't interested in being used strictly as telephone operators.
- The installation of amateur single sideband and VHF equipment is identical to that of marine SSB and VHF equipment. Since both types of systems use frequencies very close to each other, the elements of the installations are precisely the same. A ham single sideband set will use the marine SSB antenna system and frequently the same au-

tomatic antenna tuner. Only in an emergency, however, may a VHF or SSB ham set be used on marine frequencies. It is possible to modify a ham single sideband set to transmit on marine frequencies (as described in the accompanying text on " 'Unlocking' Ham Transceivers"), and the Japanese have recently introduced a 2-meter ham radio handheld that will tune in marine VHF frequencies with a slight modification. Either practice is strictly illegal except in an emergency, and a ham could lose his license if caught operating non-type-accepted (under Part 80 of the FCC regulations) equipment on marine frequencies.

- Synthesized SSB sets are now available with sufficient frequency capability to give you access to ham SSB frequencies (at 4, 7, 10, 14, and 21 MHz). When you have the appropriate ham license, and only then, you can transmit on these frequencies using your synthesized marine SSB set. Marine sets with ham radio capabilities are a good investment because the two services are almost identical in frequency, power output, and transmission mode.

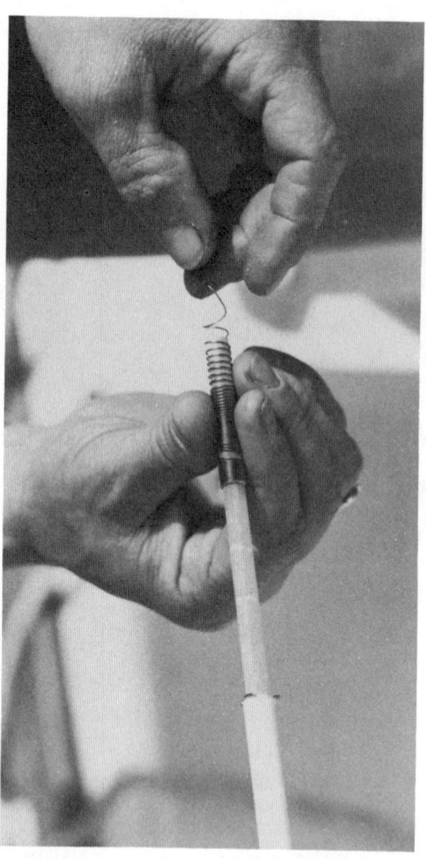

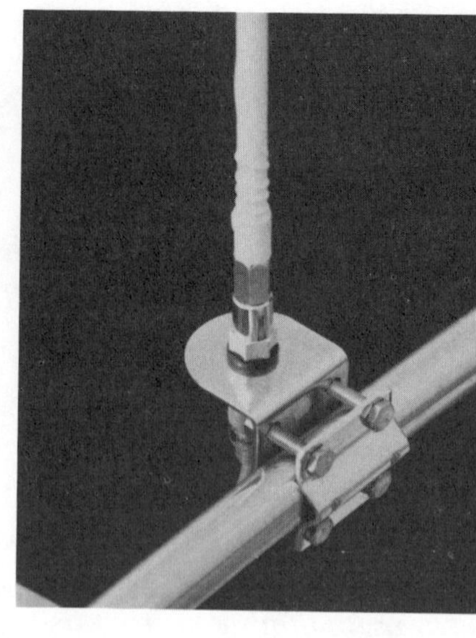

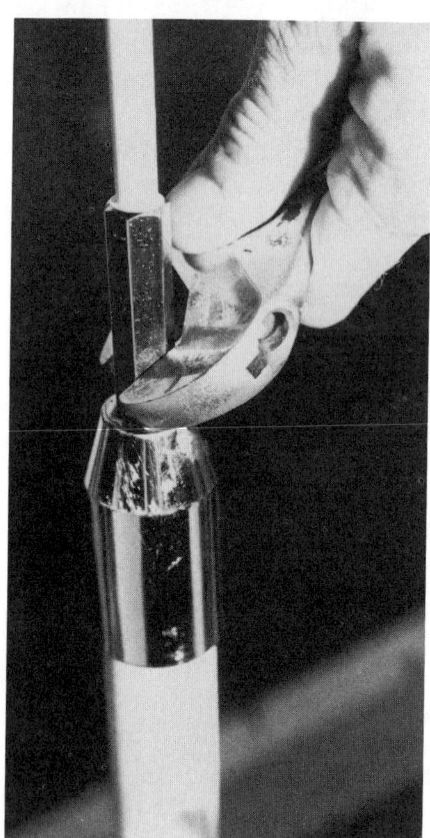

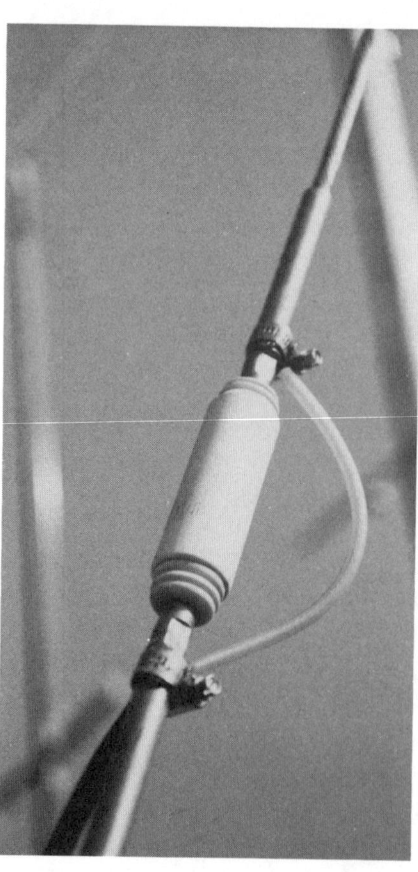

Top left. *An unusual but practical ham radio antenna system capable of working the world, any time of day or night.*
Top middle. *Ham radio whip antennas used without a tuner require retuning for marine band (as opposed to ham band) operation. This is accomplished by trimming or "pruning" the loading coil of the antenna.*
Top right. *A ham radio single-sideband whip installation. The grounded rail acts as a counterpoise. Be sure to seal the exposed coax connector.*
Far left. *Antenna maintenance includes tightening all connections once a year. Also inspect for cracked or broken fiberglass shafts; they may be mended or replaced.*
Left. *A backstay coax connection for a ham radio or marine SSB installation. Whip antennas may be pretuned to function on a single selected frequency, but a backstay antenna requires use of an antenna tuner separate from the radio.*

TIME (UTC)	FREQ. (MHZ)	NET/NAME DESIGNATOR	DAYS	AREAS
0100	3.935	GULF COAST HURRICANE NET		G/C USA
0100+	21.407	PAC-IND OCEAN NET	DAILY	PAC/IND OC
0100/2400+	14.313	MAR. MOBILE SERV. NET		PACIFIC
0200+	3.992	ARIZ TRAFFIC NET		ARIZ/BAJA
0200+	14.305	CAL-HAWAII (CC&H?) NET		CA/HAW/PAC
0200/0100+	7.290	HAWAII PM NET	M-F	HAWAII
0220+	14.315	JOHN'S WEATHER NET	DAILY	SO PAC/NORFOLK ISL
0300/0200+	14.313	SEAFARERS NET		PAC/W COAST
0300+	14.106	TRAVELER'S NET		AUST/IND OC
0330	14.040	E/C M/M CW NET		E/COAST
0400+	14.115	CANADIAN DDD NET	DAILY	PACIFIC
0400	14.075	PAC CW TRAFFIC NET	MWF	PACIFIC
0430+	14.314	PAC MAR NET-WARM UP		PACIFIC
0500+	21.200	VK/NZ/AFRICAN NET	DAILY	PAC/IND OC
0500+	14.280	USA/ AUSTRALIA TFC NET		PAC
0530**	14.303	SWEDISH MARITIME NET		PACIFIC
0530+	14.314	PACIFIC MAR NET		PACIFIC
0630+	14.180	PITCAIRN NET	MON	SO PAC
0630+	14.320/105	SO AFRICAN MAR NET	DAILY	ATL/IND OC.
0700+	14.313	INTERNATIONAL M/M NET		ATL/MED/CAR
0700+	14.265	PACIFIC ISLAND NET		PACIFIC
0700	14.310	GUAM AREA NET		WEST PAC
0715+	3.820	BAY OF ISLANDS NET	DAILY	AUS/NZ/S PAC
0800	7.280?	AUSTRAILIA TFC NET		AUS/S PAC
0800>0830+	14.315	PAC INTER-ISLAND NET	DAILY	S PAC/SEA
0800+	14.303	UK MARITIME NET		ATL/MED/CAR
0900+	14.313	MEDITERRANEAN M/M NET	DAILY	MED
0900	7.080	CANARY ISLAND NET		ATL
1000**	14.330	PAC. GUNKHOLERS NET		SO PAC
1000	14.313	GERMAN M/M NET	DAILY	ATL/MED
1030+	3.815	CARIBBEAN WX NET	DAILY	CAR
1030**	14.265	BARBADOS CRUISING NET		ATL/CAR
1100/1000+	3.770	MARITIME WX NET	M-Sa	NE CANADA
1100+	7.082.5	CARIBBEAN M/M NET	DAILY	CAR
1100+	14.313	INTERCON NET	DAILY	N/S/C AMER
1100+	14.283	CARRIBUS TFC NET	DAILY	E/C-CAR
1130+	14.320/105	SO AFRICAN M/M NET	DAILY	S ATL/IND OC.
1130+	21.325	SO ATL ROUNDTABLE	DAILY	SO ATL/IND OC.
1200	14.040	M/M CW NET		E/C USA
1200	14.332	YL EMERGENCY NET		USA
1200+	14.320	SO EAST ASIA NET	DAILY	SEA/INDONESIA/AUS
1245/1145+	7.268	WATERWAY NET	DAILY	E COAST/CAR
1300+	21.400	TRANS-ATL M/M NET	DAILY	N ATL/MED/CAR
1400	7.292	FLORIDA COAST NET		FLORIDA
1400+	3.963	SONRISA NET	DAILY	BAJA/CAL
1500**	7.193	ALASKA NET		ALASKA
1545**	14.340	MARQUESAS NET		SO PAC
1600/1500+	7.238.5	BAJA CAL MAR NET	DAILY	BAJA/CAL
1600/1700+	14.313	COAST GUARD M/M NET	M-F	ATL/CAR/USA
1630**	7.285	SERAPE NET	SUN	MEX COAST
1630+	21.350	PITCAIRN NET	FRI	SO PAC
1700+	14.340	CAL-HAWAII NET	DAILY	CAL/HAW
1700+	7.240	BAJCO M/M NET	M-F	C/AMER/PANAMA
1700+	14.313	INTERNATIONAL M/M NET	DAILY	ATL/MED/CAR
1700+	14.329	SKIPPERS NET	DAILY	PACIFIC
1730+	14.292	ALASKA NET	M-F	ALASKA
1730+	14.115	CANADIAN DDD NET	M-F	PAC
1800>1900+	14.285	KAFFEE KLATCH UN-NET	MWSa	HAW/TAHITI
1800+	14.303	UK MARITIME NET		ATL/MED/CAR
1800+	14.305	CONFUSION NET	DAILY	S/PAC
1800	14.342	GORDON ON THE AIR		W/W
1800/1700+	14.313	MAR MOBILE SERV NET	DAILY	ATL/CAR/PAC
1800+	7.076	SO PAC CRUISING NET	DAILY	SO PAC
1800	7.197	SO PAC SAILING NET	DAILY	SO PAC
1830+	14.342	MANANA M/M NET-W/UP	M-Sa	W/C-E/PAC
1900+	14.342	MANANA M/M NET	M-Sa	W/C-E/PAC
1900	7.255	WEST PACIFIC NET		W PAC
1900+	7.285	SHAMARU NET(FRIENDLY?)	DAILY	HAWAII
1900	21.390	HALO NET		N/S AMER
1900+	14.329	BAY OF ISL NET	DAILY	NZ
1900	3.855	FRIENDLY NET		HAWAII
1900	3.990	NORTHWEST MAR NET		PAC NW
2000**	7.060	VK MARITIME NET		AUS/SO PAC
2100+	14.315	TONY'S NET		NZ/SO PAC
2100	21.390	I/AMER TRAFFIC NET		N/S AMER
2130**	14.318	DAYTIME PACIFIC NET		PACIFIC
2130	14.290	E/C WATERWAY NET		E/C USA
2200+	21.350	PITCAIRN NET	TUES	SO PAC
2200+	21.404	PAC MAR NET-W/UP	M-F	PACIFIC
2230+	3.815	CARIBBEAN WX NET		CAR
2230+	21.404	15 MTR PAC MAR NET	M-F	PAC/CAL/HAW
2300/2200+	14.313	INTERCON NET	DAILY	N/S/C AMER
2310**	14.285	CAL-SO PAC NET	MON	SO PAC
2330	21.325	SO ATL ROUNDTABLE		SO ATL
2400+	14.320	S E A M/M NET	DAILY	S&W PAC/SEA

A listing of maritime mobile networks or "nets," as current as possible as of May 1986. A few nets operate in Morse code; some offer weather information; some are not primarily maritime mobile nets but may accommodate maritime mobile traffic. Stations are cautioned to check appropriate band allocations, operator privileges, and third party agreements. Note that nets on 14.313 MHz offer nearly world-wide coverage 24 hours a day. UTC = universal coordinated time, another name for Greenwich mean time. Two times separated by a slash indicate winter to summer time changes. (Courtesy Roger Krautkremer, W6SOT.)

7

Weather Facsimile Reception

by Gordon West

Weather facsimile equipment enables the mariner to inspect satellite images and charts prepared by weather bureaus, a development of immense usefulness to offshore navigators tracking weather systems in order to find favorable winds and avoid storms. There has been a dramatic recent departure from the tried and proven marine weather facsimile equipment in that you can now print weather fax charts on a shipboard computer or use a television adapter to see and store charts on a shipboard television or video system.

How It Works

"Radiofacsimile" is the descriptive term for sending pictures and maps over radio frequencies. Specific frequency tones translate as white, shades of gray, and black images when scanned electronically twice a second to produce a chart or satellite imagery. There are approximately 40 radiofacsimile weather broadcasting stations throughout the world, and they transmit on shortwave frequencies that may be tuned in with any quality shortwave receiver or general coverage ham set. Although areas of coverage and types of transmitted weather information vary, almost all stations offer the following:

- Satellite imagery
- Surface weather analysis and pressure systems
- Surface weather prognosis for 24 and 36 hours
- Extended surface prognosis for up to 5 days
- Ocean wave analysis, including height and direction
- Ocean wave prognosis
- Iceberg analysis, or water temperature zones

On the Pacific Coast, weather facsimile stations are more likely to offer water temperature charts than iceberg analyses. On the Atlantic Coast, you can get both.

In the United States, weather charts are prepared (handwritten) by either the Navy or the National Weather Service. The Navy charts are complex and detailed, and it takes a trained weather map reader to receive their full value. The National Weather Service tends to generalize on their charts, making them as readable as possible for mariners that may not have a long background in chart interpretation.

The high-powered transmitting stations are usually government supported—by the Coast Guard,

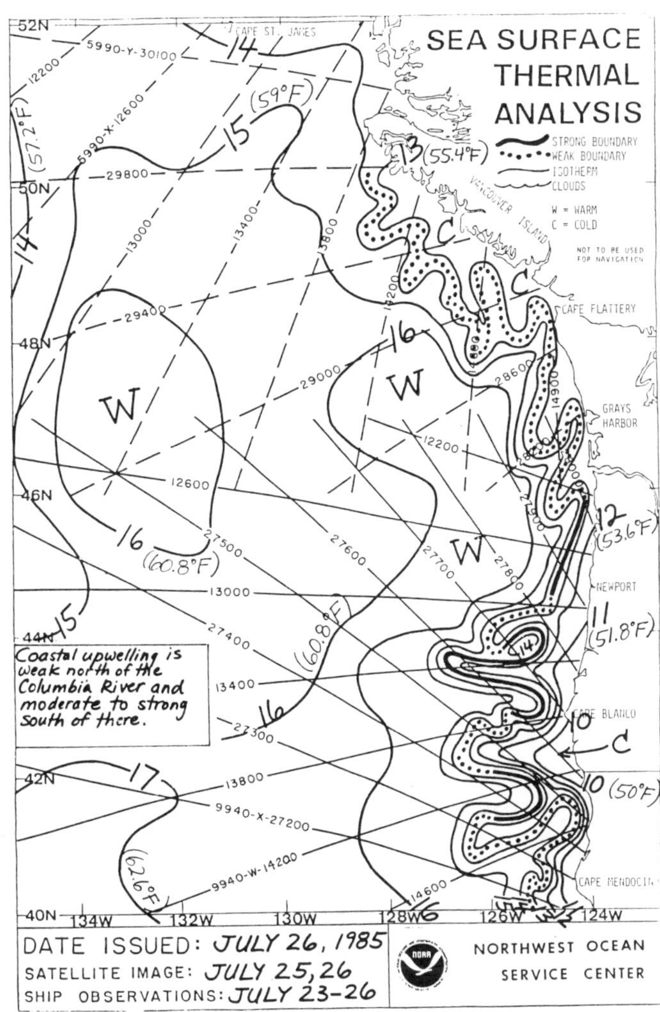

Information on water temperature is transmitted by weather facsimile stations.

Air Force, or Navy in the United States. A few stations are privately funded in connection with the National Weather Service. Each station broadcasts its weather charts on a specific schedule, but, unfortunately, broadcasts are often late. On manual machines, a lot of paper can be wasted waiting for a weather chart to come on line. The charts are transmitted simultaneously on several different frequency bands, some of which, such as 4 MHz, 6MHz, and 8 MHz, propagate better at night than during the day, while others, including 12 MHz, 16 MHz, and 22 MHz, propagate better during the day. A mariner has only to tune in the band that offers the best reception for his particular area of cruising. Within 1,000 miles of the transmitting station, 4 or 8 MHz might be the best bet, while

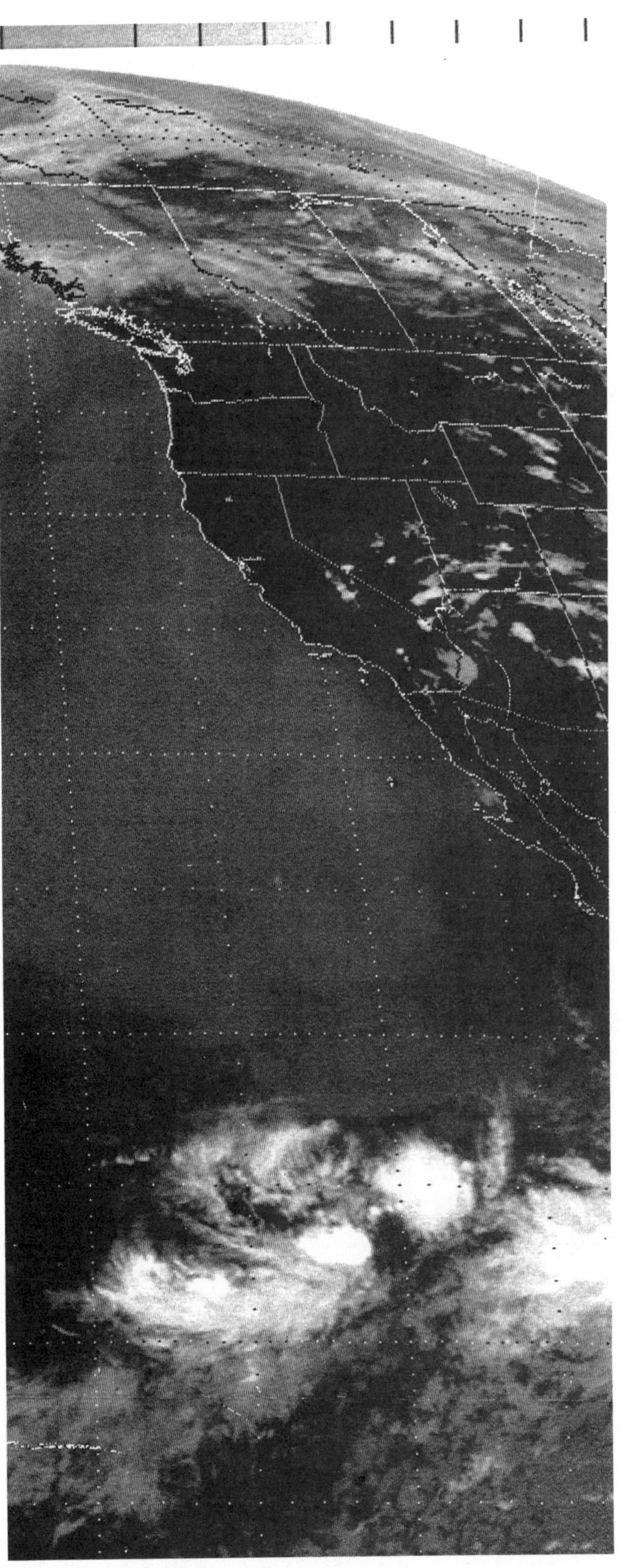

Satellite imagery from a weather fax. The portion reproduced here is at full size.

3,000 miles away and in the daytime, the 12 or 16 MHz band is the better choice.

The weather charts are updated every 6 hours, and satellite images are updated whenever new pictures are available. Due to the loss of several weather satellites during launch, the number of satellite pictures may be limited to only two or three per day.

You can receive a list of weather facsimile transmitting stations by writing Alden Electronics, Washington Street, Westborough, Massachusetts 01581 (Phone (617) 366-8851), and if you are really interested in weather forecasting, you may wish to order their radiofacsimile book for $12.45. Alden, one of approximately 10 manufacturers of weather facsimile equipment, has been a pioneer in the field, and they have contributed greatly to the advancement of weather facsimile for mariners.

What's Available

The shipboard weather facsimile installation consists of a 3 to 30 MHz shortwave receiver with single sideband capabilities, an antenna and ground system, and a printer. The more advanced setups feature a specialized receiver that scans for the best band of reception, automatically activates and stops the recorder when chart broadcasts are started and stopped, and can be programmed to receive weather broadcasts from stations on different bands at different times of day. Thanks to smart, microprocessor-based receivers, all of this scanning, starting, stopping, and prioritizing is done electronically with little additional cost to the customer.

The traditional weather facsimile system contains the receiver and printer in one housing. As an alternative, for the mariner who already possesses a good single sideband receiver, there are separate rotating drum recorders available. Drum recorders, which are also used in the traditional weather facsimile packages, first used wet paper to "burn" their images, but the paper has a tendency to dry out, and year-old, sealed paper is simply not the same as brand-new wet paper. It also costs at least $25 per roll, and today there are almost no wet paper machines left. The papers now used in drum rollers won't dry out, and they give high-resolution prints. These papers may be illuminized, electrosensitive, or thermosensitive, each type having its strong points. Illuminized gives the sharpest picture but is the most expensive. Thermo and electrosensitive papers give good readouts and cost less per roll. A single weather chart requires approximately 8 inches of paper and 10 minutes, with the unit scanning at 120 times per minute. The process is similar to the stylus depthsounder, in which the rotating stylus will eventually print out a picture that is recognizable.

One new manufacturer, Latitude, offers a weather facsimile module that plugs into a high-resolution computer printer and any high-quality single sideband receiver, thus superseding the drum recorder. In this system, a home computer's high-resolution printer serves double duty by rapidly printing out the charts after the weather service completes its transmission. The Latitude module stores the information, and then quickly dumps it out to the computer, ready for the next chart to be broadcast. It also has provisions for starting and stopping the computer when the chart starts and stops. Resolution is surprisingly good, although quality is dramatically improved by going to the very best printer available. The charts are usually smaller than what you would receive from a drum recorder, but they are just as legible.

At the same boat show in which Latitude introduced its computer printer—fax interface, a company called Sea Lutions introduced a module that plugs into a top-quality single sideband receiver and a computer video monitor. The result is a paperless system that can recall up to four stored weather facsimile pictures. Its features include automatic phasing and locking onto weather facsimile pictures and a battery backup circuit that keeps the four pictures in memory even if the ship's 12 volts to the unit is interrupted.

You can also view weather charts on a standard television using any Radio Shack RF modulation system. Resolution will be inferior to that of a computer monitor, but you nonetheless can see the chart well enough to read it.

A shortwave receiver is a less expensive, receive-only alternative to a ham radio or SSB transceiver. It will receive weather facsimile broadcasts if, as in the shortwave receiver sections of the dedicated weather facsimile sets manufactured by Alden and others, it has single sideband capabilities. As men-

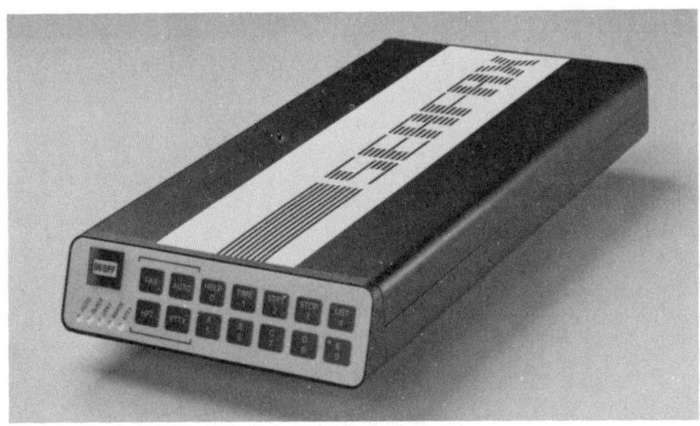

A weather fax converter will link a shortwave receiver to a computer printer, providing charts similar to the sea surface thermal analysis shown earlier.

tioned, the shortwave receiver may be built into the weather facsimile housing or housed as a separate module. Ham radio manufacturers such as ICOM, Kenwood, and Yaesu offer high-quality non-portable shortwave receivers that will tune in and lock onto weather facsimile signals in the single sideband mode. The Sony 2010 portable shortwave receiver can also be used with good results, but Sony is about the only manufacturer of a portable shortwave set that offers single sideband reception. Like a ham or SSB transceiver, a shortwave set needs to be wired to a recorder or computer printer to produce a chart.

The antenna system need be nothing more than a fiberglass whip or even a long piece of wire run up in the rigging. Wire with a plastic insulating sleeve should pick up lower as well as higher bands. Weather facsimile antennas that are tuned broadband are also available from many manufacturers; Metz leads the pack with their stainless steel antenna, which can be stowed out of sight. All antenna installations require a good RF ground for maximum reception on SSB frequencies. The principles of grounding are discussed in the Single Sideband chapter.

8

Radio Direction Finders

by Gordon West

Just below the AM broadcast band are frequencies allocated for radio navigation. The U.S. Coast Guard operates marine radio beacon stations that dot the coastline and islands in all U.S. cruising areas, and every major boating area and seaport worldwide maintains similar stations. In many areas of the world where loran signals are inaccurate or nonexistent (such as most of the Caribbean and the waters between Baja California and Panama), the radio direction-finding system provides a means of position finding and of homing into ports. These "point source" radio beacons transmit on frequencies between 190 and 410 kHz. All marine radio beacons are indicated on marine charts, and frequencies and broadcast schedules are given in the Defense Mapping Agency's publications No. 117A and No. 117B, *Radio Navigational Aids*. Some local marine electronics dealers give away frequency lists that include local radio beacons. Aeronautical radio beacons operated by the Federal Aviation Administration can also be used for direction finding; these are sometimes located on marine charts, or you can pencil in their precise locations if you know them. Aeronautical charts will show the locations of these beacons. Finally, commercial AM broadcast stations (550 kHz to 990 kHz) may be used for direction-finding purposes if the transmitters are advantageously located.

Two types of radio direction finders are available for tuning in "point source" stations on radio beacon and AM broadcast band frequencies. These are the portable or handheld radio direction finder (RDF), which can cost as little as $200 or as much as $400 or more, and the automatic direction finder (ADF), which will cost between $1,000 and $1,500. Both tune the same frequencies, but the ADF "fixes" on the point source station automatically, while the RDF requires you to do this manually.

With portable radio direction finders available for about $200, it makes sense to have one of the small units aboard even if you depend primarily on more sophisticated electronic navigational systems such as loran and SatNav. The latter have complex microprocessor circuitry and external antennas mounted in the weather, and on both counts they are vulnerable in ways that an inexpensive RDF set is not. And with its own batteries, a portable RDF will function even when the ship's power has been lost. There is no plan to phase out the radio direction-finding system; the Coast Guard will continue to maintain their radio beacons for the foreseeable future. Loran signals can become less predictable

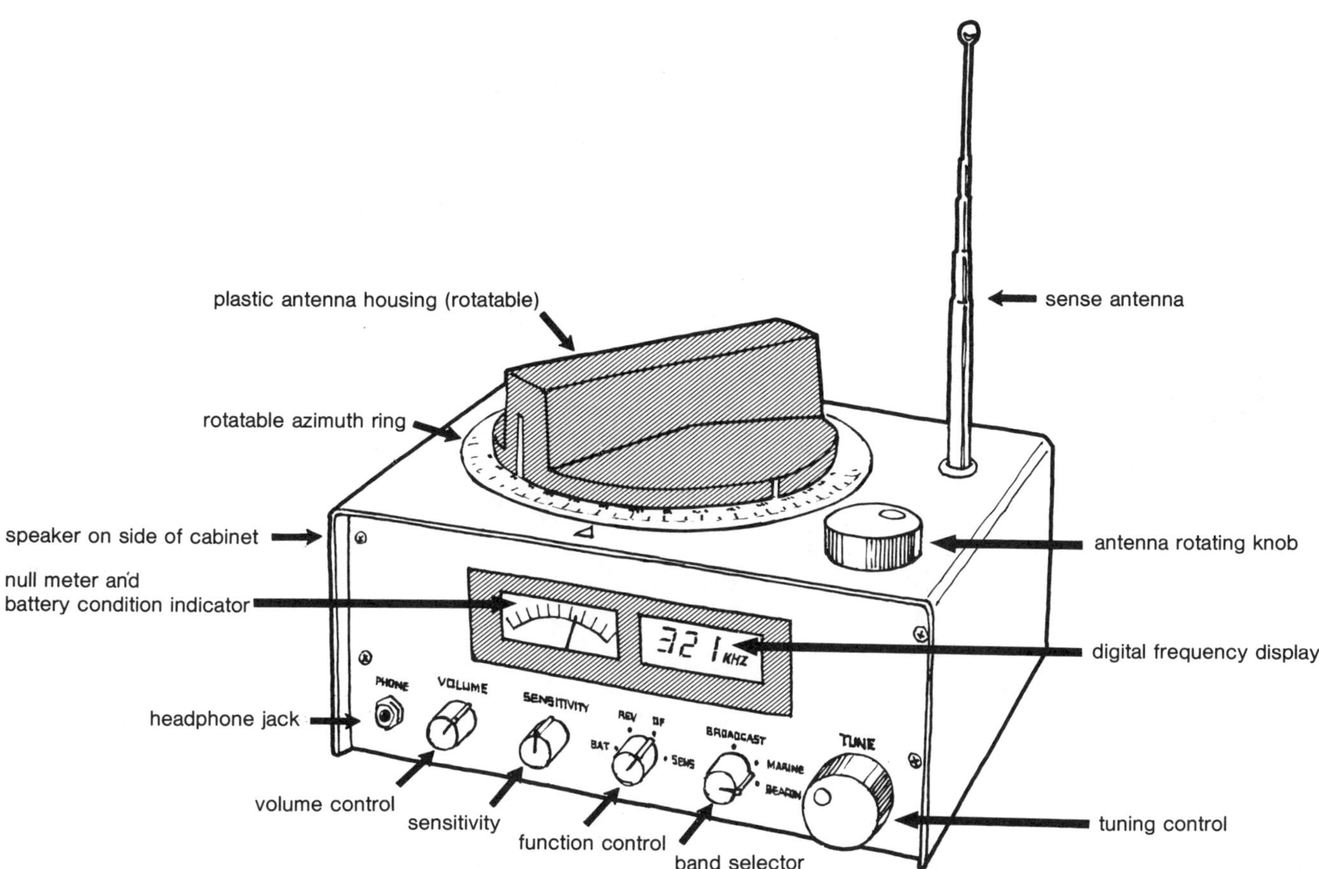

Typical portable RDF covering the radio beacon and AM broadcast bands. Some makes also have a beat frequency oscillator control, which helps in tuning a weak station by providing a squeal when the BFO is on to aid in finding the null. Others may include the 2-to-3 MHz marine band, but it is of little value for navigational purposes. The function control knob's four positions enable the user to check the voltage of the built-in batteries, to choose straight reception of AM broadcast stations (without any direction finding), to use beacon or AM stations for direction finding, or to energize the sense antenna. The price of a unit like this ranges from $200 to $500, and a typical size is 10 inches wide, 12 inches front to back, and 6 inches high.

inside harbors and close to jetties, situations in which (provided the harbor or jetty is provided with a radiobeacon) an RDF is at its best.

How It Works

The U.S. Coast Guard operates two types of radio beacons—marker beacons that broadcast low-power signals from local harbors, and more powerful radio beacons that mark headlands (these are often located in lighthouses), major harbor entrances, drilling rigs, and other such features. The normal range for low-power beacons is 10 miles, while the high-power beacons can transmit their signals over distances of up to 150 miles, making them useful for position fixing by triangulation. High-power beacons are sometimes found in groups of up to six, transmitting sequentially in a timed pattern. Both types transmit very slow Morse code identifiers, making it easy for you to determine which beacon you have tuned in.

Radio waves between 200 kHz and 900 kHz (which includes the commercial AM broadcast stations as well as marine radio beacons) normally travel in straight lines radiating outward from the transmitter. In theory, all you need do is pick up the signal, determine what direction it's coming from, identify its source, and find the location of that source on your chart, and you'll have a line of position (LOP) along which you must be located.

The heart of a radio direction finder is its specialized antenna, which is highly directional to incoming radio waves. On a portable RDF the antenna windings are housed in a plastic bar that turns freely atop the receiver. Reception is strong when the bar is broadside to the incoming signal path, but when you swing the antenna end-on toward the signal source, the signal will disappear (a null). Because it can be located with much more precision than can the position of strongest signal reception, the null is used for direction finding. The unit is usually equipped with a meter to help locate the null. When it is located, the bearing is read on the 360° compass rose under the antenna. Either end of the antenna will elicit a null, so the user must make an independent determination as to which of the two possible directions (180° separated) the signal is coming from. This should rarely be difficult. If you are navigating on the West Coast, for example, a shore station signal is bound to be toward your east.

In the automatic direction finder (ADF) the principle is the same, but the antenna is a fixed double loop mounted outside the pilothouse, and signals are electronically "nulled out" while the twin loop assembly remains stationary. A pointer or digital readout will display the bearing of a signal. The ADF antenna also frequently incorporates an external sense antenna to resolve that 180° bearing ambiguity.

The procedures for using a portable RDF are simple. Let's say you want to home into a low-power radio beacon transmitting from a harbor. Swing your vessel into the approximate direction of that harbor, then tune in your RDF to the frequency listed on the chart. Doublecheck that the identifier you hear is indeed the marker beacon on the harbor entrance. Now swing the loop and look for the null. If the signal disappears precisely over the bow, you are heading toward the harbor, but if the signal disappears on the starboard bow, you need to alter course to starboard in order to home directly on the harbor entrance. If after a half hour of cruising on a steady course the radio beacon signal again appears a few degrees on the starboard bow, you are probably being set by wind or cur-

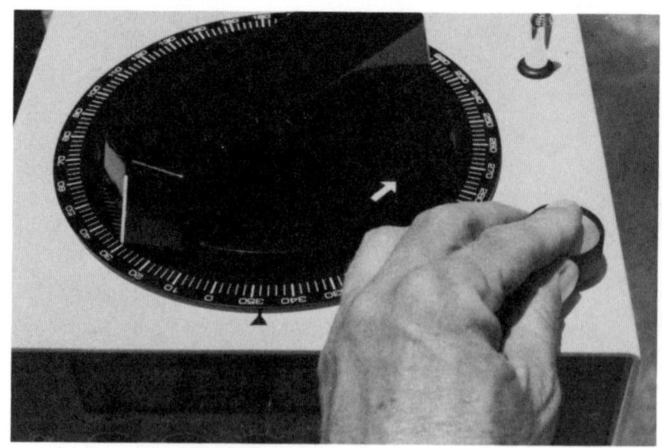

Rotating the antenna on a portable RDF.

rent. In this way an RDF can tell you how wind and current are affecting the accuracy of your dead reckoning. Obviously, these procedures will work just as well when you want to run a course directly *away from* a radio beacon, adjusting course to bring the "null" directly over the stern.

One of the most important uses for an RDF or ADF is for obtaining position fixes. Two or more bearings on chart-plotted point sources are needed for a fix, and radio beacons or commercial broadcast transmitters will enable you to get those bearings in low-visibility conditions, when you need them the most. Set the rotatable azimuth ring on your RDF to 000 in order to read the bearings in degrees relative to the ship's course, or rotate the ring to your current magnetic heading and read the absolute magnetic bearings. Try to find at least three stations, which should be in different compass quadrants; the wider the crossing angles of the bearings, the more accurate your fix. The stations might include a high-power beacon transmitting from a distant harbor on your port bow, a strong AM radio station with a shoreside transmitter on your starboard beam, and a small jetty marker directly astern of you. If the bearings are measured relative to the vessel's heading, convert them to absolute bearings in degrees magnetic. Plot the three lines of position (LOPs) from the charted transmitter locations. The LOPs should form a small triangle, the familiar navigator's "cocked hat." Your probable position is within this triangle.

There is also a simple radio direction-finding technique for checking speed over the ground. Find two shoreside radio beacons and on your chart draw a line from each to intersect your course heading at 90°. Now measure the distance along your course between the two intersecting lines. Suppose that distance is 11 nautical miles. Start the clock when you are exactly abeam of the first beacon, and stop it when you are abeam of the second. If two hours has elapsed, your speed is about 5½ knots.

The accuracy of RDF bearings is seldom better than 2 degrees one side or the other, and sometimes it can be worse. When stations transmit signals that must bend around landmasses or travel over hills or through valleys, the radio waves are likely to be redirected before they reach the ocean. For this reason it is important to know when the signals you are receiving from a radio beacon or AM broadcast station have traveled significant distances over an intervening landmass. These signals may have been redirected by as much as 20 degrees. This is one reason why marine charts don't list inland aeronautical radio beacons or AM broadcast stations for direction-finding purposes.

Another characteristic of radio waves just below the AM broadcast band is that they may bounce off the ionosphere (an envelope of ionized gases above the earth) at nighttime, sometimes giving you highly erratic bearings to the point source beacon. These "skywaves" may approach from a direction as much as 35 degrees away from where the station is transmitting. Be extremely cautious of nighttime bearings on distant high-power radio beacons and AM broadcast transmitters. Within 40 miles of the transmitting station you are not likely to have any problem, but beyond that distance, groundwave signals may be overridden by skywaves, and this will give you a false bearing.

Despite the inaccuracies created by skywaves and coastal refraction of signals, a tremendous advantage of RDFs over other radio positioning devices is their simplicity of use. When you approach a fogbound harbor that has a radio beacon at its entrance, the accuracy of the bearings improves as the intervening distance shrinks. By the time you are within a few miles of the beacon, your positional error can be measured in yards. Furthermore, since you are homing on the beacon, it is only necessary to steer a course that nulls the beacon when the antenna is pointing dead ahead. No need to plot the displayed latitude and longitude on a chart, as you would need to do with a loran receiver, and from that fix, which could have considerable error in it, replot a course to the harbor.

Other, sometimes major inaccuracies can be introduced after the incoming signals reach your boat, but these you can control with the proper precautions. Metal surfaces and objects, for example, will deflect and distort radio waves. With this in mind, keep your RDF or ADF at least two feet above metal structures when direction finding, or better (since you can't very well operate your receiver two feet above the metal mast and rigging), develop a deviation card for the receiver in its normal position aboard the boat. The procedure is this: On a clear day, take a series of visual bearings on a local radio beacon (one on a jetty would be good) using a pelorus or a hand-bearing compass, and compare each such bearing to the corresponding RDF or ADF bearing. Three people—one on helm to swing the vessel, one to call off visual bearings, and one to take the RDF or ADF bearings—make the operation smooth. If there are major deviations, you may need to find a new location for the receiver. Metal stanchions can cause deviations of up to 10 degrees, and even the aluminum frame of a pilothouse window can redirect incoming radio waves. When you find a suitable and convenient spot, develop the deviation card for that spot. The operation is analogous to "swinging a compass."

When you use a portable RDF, which should be sitting on a tabletop or other convenient horizontal surface, make sure the set's lubber line is parallel to the vessel's fore-and-aft centerline. Naturally this applies equally to a permanently installed ADF. Just as important, make sure the helmsman holds a steady course while you take the bearings. Both these precautions are critical to the accuracy of the bearings.

You may need to shut down your engine to pick up weaker signals. The most common sources of interference include fluorescent lights, engine ignition systems, alternator whine, and electric motors.

What's Available (And Buying Tips)

There have been few advances in the technology of radio direction-finding equipment over the last few

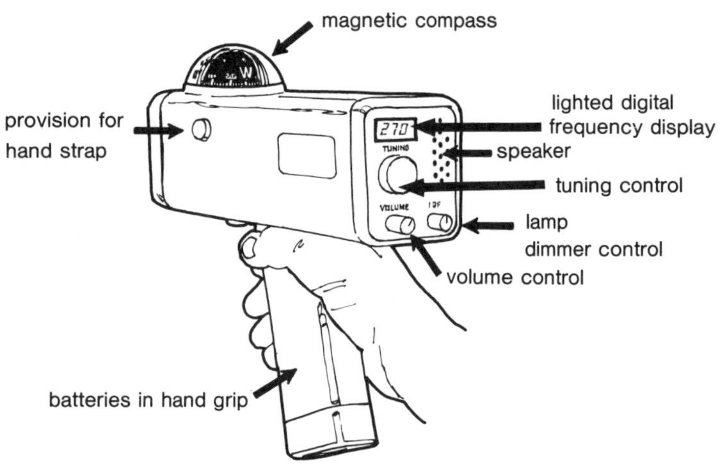

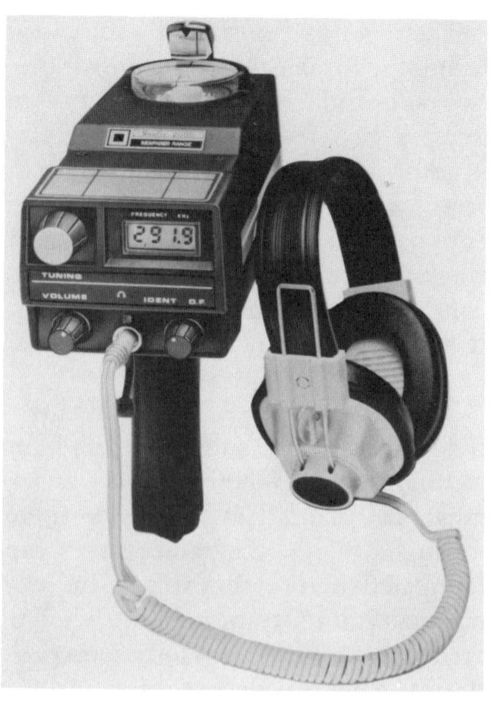

Headphones on a handheld RDF aid the null determination.

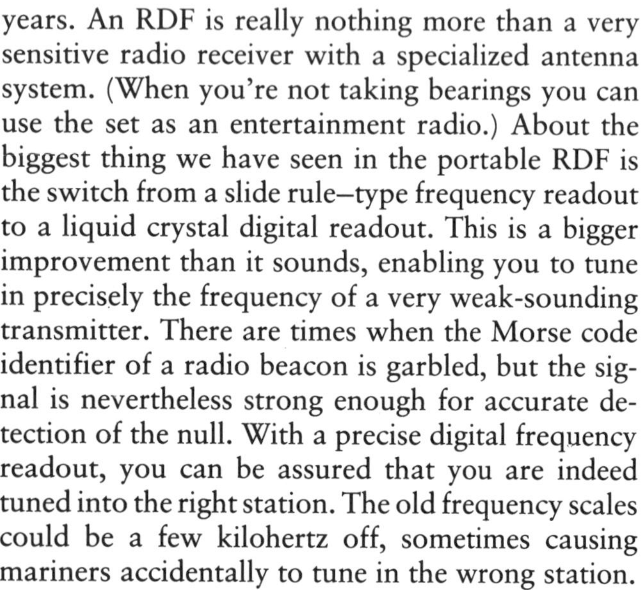

Handheld RDFs come in a variety of packages, but essentially they have a pistol-type grip, with the antenna and receiver controls above the grip and a small compass mounted above the antenna section. Some have a locking button that can be pressed when the null is audibly observed, which locks the compass card at that position. This is a useful feature when the boat is rolling or pitching heavily, or at night when the compass card may be hard to read due to its small size. The price of a handheld RDF ranges from $300 to $800.

years. An RDF is really nothing more than a very sensitive radio receiver with a specialized antenna system. (When you're not taking bearings you can use the set as an entertainment radio.) About the biggest thing we have seen in the portable RDF is the switch from a slide rule–type frequency readout to a liquid crystal digital readout. This is a bigger improvement than it sounds, enabling you to tune in precisely the frequency of a very weak-sounding transmitter. There are times when the Morse code identifier of a radio beacon is garbled, but the signal is nevertheless strong enough for accurate detection of the null. With a precise digital frequency readout, you can be assured that you are indeed tuned into the right station. The old frequency scales could be a few kilohertz off, sometimes causing mariners accidentally to tune in the wrong station.

Most portable RDFs operate off flashlight batteries, and none are weatherproof. Keep RDFs and ADFs out of the moisture.

Some portable RDFs have an auxiliary "sense" antenna, the function of which is to resolve that 180-degree ambiguity. Given an accurate frequency tuner, an audible Morse code signal, and a local chart, however, you should have no problem deciding which of the two possible directions is the actual one. If the sense antenna comes on the unit you want, fine, but its presence or absence need not be a criterion of selection.

Handheld RDFs are available, these differing from the portable sets in that the antenna is housed in a pistol grip and the entire unit is rotated for null determination. A magnetic compass is substituted for the azimuth ring, which makes the unit easy to use but also introduces a new source of inaccuracy: The compass is of course subject to magnetic deviation, and the small size of the compass card limits the precision of the reading. The handheld RDFs have their chief application aboard sailboats, since they can easily be taken on deck and held outboard.

away from the standing rigging. They're priced in the same range as the portable units.

Most of the characteristics to look for in an RDF are similar to those that apply to VHF radios, as discussed in Chapter 3. An inexpensive RDF offering CB and FM bands (useless for direction finding) as well as the radio beacon and AM broadcast bands should be viewed with suspicion. Its "sensitivity" may be low, its "selectivity" poor, and its null may not be sharp enough for accurate bearings.

Sensitivity, the ability to pick up weak signals, is expressed as microvolts per meter, a smaller value meaning a greater sensitivity. The value should probably be 50μV/m or less on the radio beacon band, 40μV/m or less on the AM broadcast band. Selectivity is the ability to receive the desired station while filtering out interference from other stations, perhaps with stronger signals, on adjacent frequencies. Selectivity is rated in decibels—the higher the value, the better the selectivity. The value should probably be 30dB or higher. As with VHF radio, actually trying out the sets side by side in a marine electronics showroom is preferable to depending on quoted specs.

Automatic direction finders *can* be portable: The antenna is housed in a bar atop the set, as in the portable RDF, but a motor rotates the antenna automatically. Performance of these units is compromised by the motor, which can drown out the signal.

A permanently installed ADF with its large outside double-loop antenna can give you greater direction-finding range, many times greater than that afforded by a portable RDF. These units should be installed by a technician, since proper installation of the loop antenna's 8- to 10-strand wire is difficult, and because the units need to be calibrated when they are installed. Proper calibration increases the accuracy of the units, but again, a deviation card is necessary. These ADF units operate off the ship's batteries, so if you lose the ship's power, you lose the ADF. For this reason, and because a good ADF will cost considerably more than a portable RDF, an RDF is the more appropriate choice for an inexpensive alternative or a backup to loran.

A couple of additional things to consider when you shop for a radio direction finder:

Look for liquid crystal displays on RDFs and ADFs. The more traditional mechanical pointers tend not to be as precise or as troublefree. LCD readouts for both frequency and direction can give remarkable precision.

Look for an auxiliary connection on a portable RDF to take an external 12-volt supply. This conserves the unit's internal batteries.

Look for a light on the RDF for night viewing.

An RDF is a prime candidate for catalog purchase. Almost all units are imported, and the catalogs usually offer far lower prices than marine electronics dealers. Most dealers won't work on the insides of defective RDFs, so why not buy through a catalog and ship the set back to the factory if it needs service?

If you buy an ADF and have it dealer-installed as part of the sales package, insist that it work properly and is properly calibrated, even if that means calling the technician back for one or more additional visits.

Maintenance and Troubleshooting

Always remove the batteries from a handheld or portable RDF before storing it. Old or dead batteries could begin to leak, damaging the battery compartment.

The biggest problem with portable RDFs is intermittent contact in the antenna assembly. The antenna windings housed within the plastic bar spin freely on tiny contacts that transfer the incoming signals from the antenna to the receiver section. These contacts usually develop small amounts of corrosion after a year or so of little use. The usual rotatable loop may be pulled off the RDF set and the contacts cleaned using television tuner cleaner (GC Electronics Tuner Tonic). After the contacts have been sprayed with tuner cleaner, the antenna is reassembled and spun many times.

VHF ADF

VHF automatic direction finders are the latest rage among fishermen because they can determine the direction of any incoming marine VHF transmission. If a commercial or sportfisherman locates fish and that information is disseminated over VHF

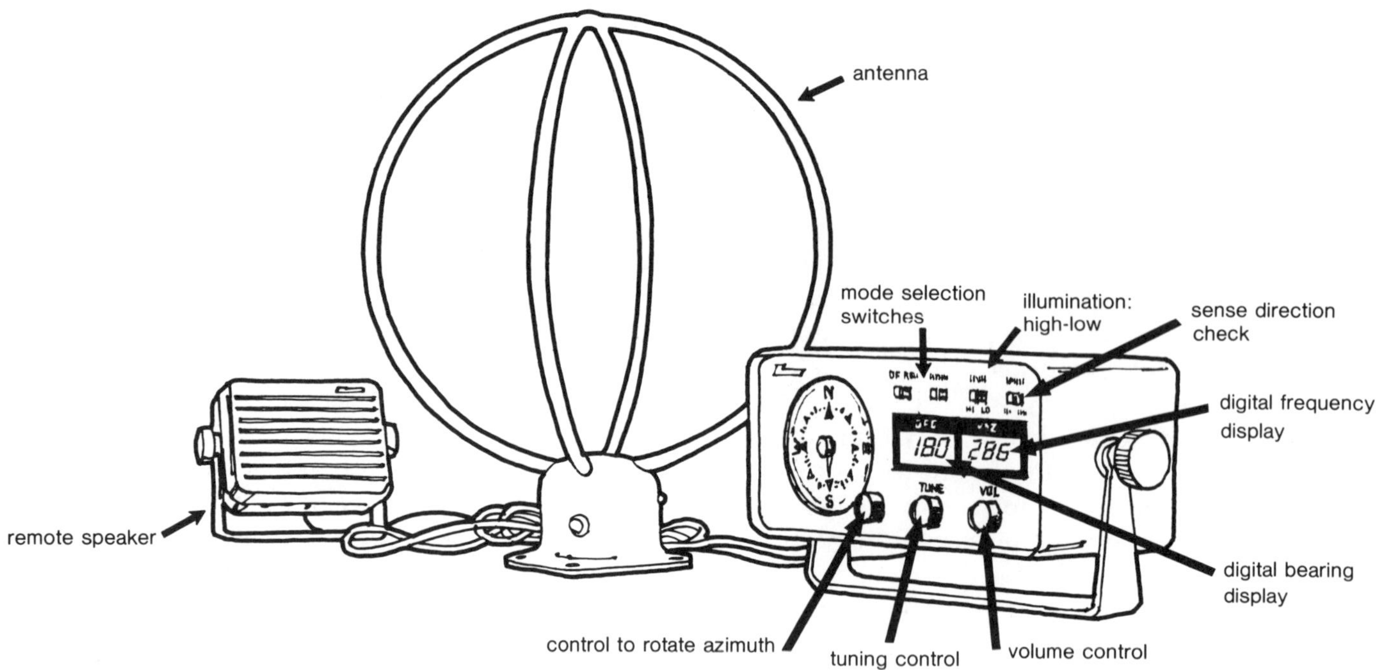

The two elements of an automatic radio direction finder are the topside double-loop antenna and the control and display console. The speaker may or may not be built into the console, depending upon make. The circular azimuth at the left side of the console can be rotated to the course being steered to provide magnetic bearings from the pointer over its face, or it can be aligned "north-up" to display relative bearings. There are also digital displays of bearing and frequency.

Most ADFs include both marine radio beacon and AM broadcast bands; some also include the 2-to-3 MHz band, which is, however, of little use for navigational purposes. The price range of a unit like this is $2,000 to $4,000 with antenna, and a typical console size is 12 inches wide, 10 inches front to back, and 12 inches high. The antenna is typically 3 feet high.

channels, he may soon have company. VHF automatic direction finders are easily spotted by their unique four-pole "Adcock" antenna system, which is comprised of four halfwave vertical elements electronically phased for making bearing determinations. Incoming signals are received by the four-bay antenna system and electronically mixed and matched, and the resulting bearing is displayed by a pointer needle, flashing light-emitting diodes, or a combination arrow and digital readout. If the antenna system is properly mounted high above all other antennas, bearing accuracy on VHF frequencies is usually better than 5 degrees. The units are obviously useful for search-and-rescue operations.

VHF signals from other ship stations cannot usually be received much more than 15 miles away.

This is because VHF radio waves don't bend over the horizon, so that ships must be nearly within line of sight before VHF signals from one can be received by the other. This is why height of the VHF ADF antenna setup is very important.

Since some shoreside VHF stations operated by the Coast Guard or telephone companies have lofty perches, you can often tune in land stations over much greater distances, but few shore stations tell us where their transmitting antennas are actually located. A U.S. Coast Guard station in a particular harbor may use five or six transmitting antennas covering as much as 100 miles of coastline, and the local marine operator for your port might be transmitting from an antenna 75 miles outside the harbor limits. These remote antenna sites can

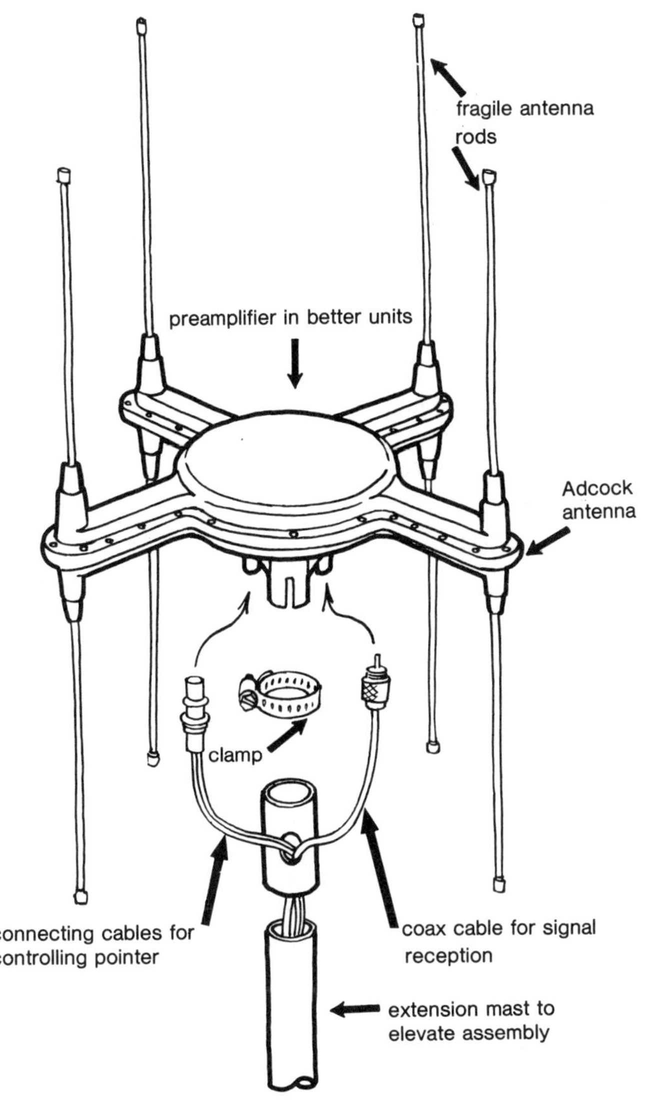

The antenna assembly used with VHF ADFs must be mounted above the boat's superstructure and away from other antennas. Alignment of the four antenna elements is critical, but easily done before securing the assembly to the plastic supporting mast.

A VHF ADF antenna, slightly askew, mounted as high as possible for good range. Below it are a radar antenna housing and a radar reflector.

complicate the use of VHF ADF for direction finding. These sets do *not* have the frequency capability to receive signals from radio beacons or AM broadcast stations. They are highly specialized for homing on other vessels.

Since VHF radio waves reflect quite nicely off buildings, billboards, metal masts, rigging, and even nearby VHF antennas, it may be impossible to home accurately on a distant ship station if you are in a harbor. You might "see" a ship station signal coming from shore rather than from sea. Once at sea, bearings will begin to settle down.

You can estimate the distance to a ship station by his signal strength on your VHF ADF set. If the signal is weak and plagued by static, the station is probably more than 10 miles away. If it's loud and clear, the station is within 5 miles.

The selection of VHF ADFs is large. For $1,000 you can buy one already built into a 78-channel synthesized 25-watt marine VHF set. The set operates like a normal VHF radio until you turn on the ADF portion, at which point it switches from the big transmit and receive antenna to the four-bay ADF antenna. The ADF antenna doesn't have the gain of a big white fiberglass whip, so reception will decrease slightly.

Another alternative is adding a VHF ADF assembly to your present VHF set. Regardless of what type or brand of VHF set you may already own, it's very easy to add VHF automatic direction-finding capabilities to it. These "add-on" devices normally cost around $500, and they are just as accurate as the built-in units. Audio from your present VHF set is taken out of the rear speaker jack and fed into your new ADF companion unit. The unit processes the audio through a phase comparison system while electronically "swinging" the four-bay antenna. It's a sensible way to go if you already have a good, synthesized, 25-watt marine VHF set installed.

These VHF ADF add-on units can also be tied into marine VHF handheld sets. The audio out of the handheld external speaker jack is fed into the ADF unit.

It's critical that the four-bay antenna be mounted well above all other antennas on your boat. Putting it down low will only cause incoming signals to distort, reflect, and refract. Some mariners might consider that weird antenna mounted high above all others on their boat a visual drawback.

When shopping for a VHF ADF inspect the antenna assembly. Are the rods hollow or solid? The hollow antenna elements will break easily, rendering your unit temporarily useless. The solid rods, such as those found on some of the newer units, will take more abuse and will bend rather than break. Be sure too that an extension mast is included with the antenna.

9

Loran

by Gordon West

Imagine yourself groping through the fog without benefit of radar, trying to get back to port with only a few gallons of fuel remaining. For the last four hours there has been no indication of land, no break in the weather—then suddenly, you pass a signpost firmly planted in the water that tells you where you are, which way to head for port, and the precise distance remaining along with an estimated time of arrival. Sound impossible? The signpost stuck in the bottom is a little farfetched, but a very sophisticated (and not so expensive) loran receiver will give you the same "waypoint" information. This type of capability is available in a small box that draws no more current than a running light.

"Loran" stands for long range aid to navigation. The system has been around since World War II, but not until the last 10 years has it enjoyed much popularity among pleasure-craft users. Loran manufacturers outside the United States have outdone themselves in producing smart loran receivers—and prices continue to fall.

In 1974, the U.S. government inaugurated loran-C as the primary civil navigation system for the U.S. coastal confluence zone, covering navigable waters from the shoreline out to 50 miles offshore (or to the 100-fathom contour where this is beyond the 50-mile limit). The Great Lakes are also covered. Most loran signals travel well beyond the coastal confluence zone and can be used at dis-

tances up to 1,200 miles from the loran transmitting chain under some conditions.

Concurrently with the expansion and fine tuning of loran stations in the United Stations, Canada began improving loran-C coverage on its east and west coasts. There is, as well, extensive loran coverage in the North Atlantic, Europe, the Mediterranean, Japan, Alaska, and Hawaii.

Loran makes possible precise and reliable position fixes. A typical fix in a strong signal area is accurate within a quarter mile, while on the fringes of loran coverage, the average accuracy is within 2 to 5 miles. The repeatability or precision of a fix is much better; you can return to within 50 feet of the exact spot where you lost an expensive anchor (or a person) overboard. It is this repeatability characteristic that enables bottom trawlers to avoid snags that have torn their nets in the past, lobstermen to find their buoys in the fog, and mariners to find in foul weather the navigation buoy whose loran coordinates they recorded when the weather was fair.

How It Works

We'll start with an oversimplified example: a man rowing his boat in the middle of a large harbor. At

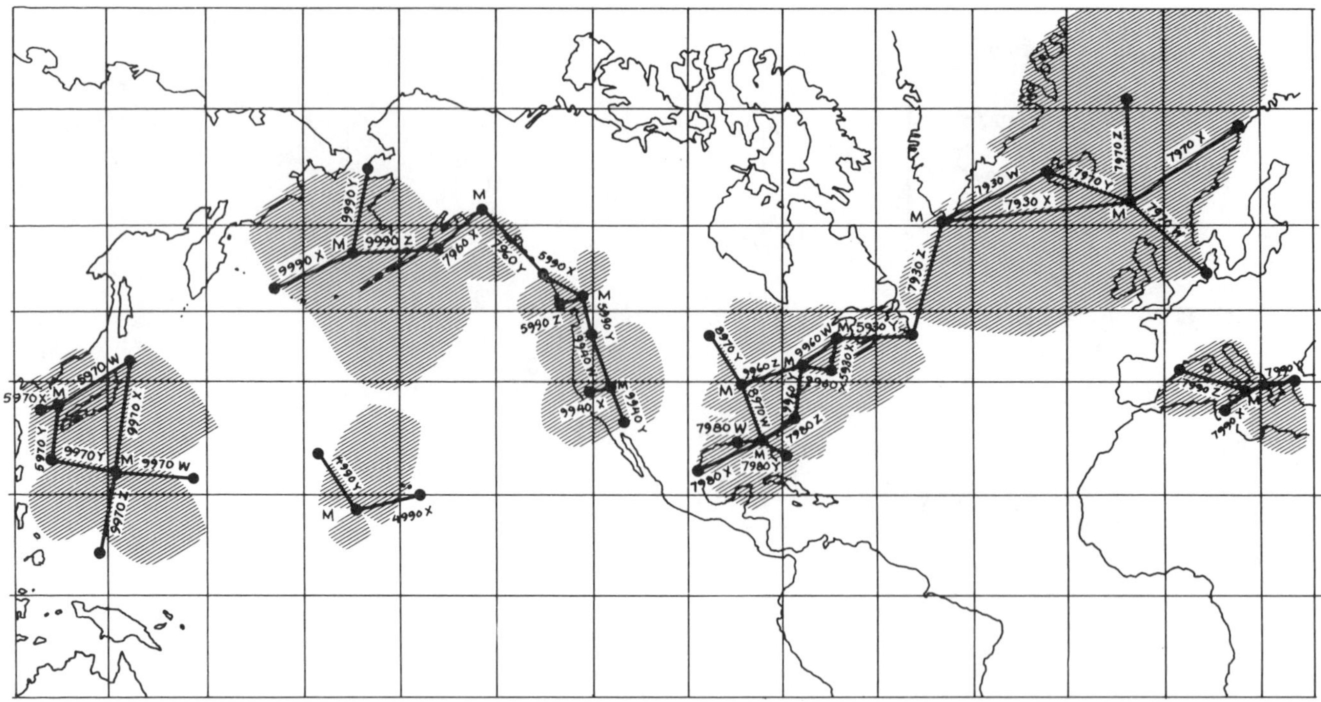

Groundwave coverage of the loran chains is denoted by the shaded areas. Ironically, the greatest coverage is in the extreme North Atlantic, where it is of little value to the average pleasureboat owner. The nominal seaward limit of the coverage as shown here is seldom realized.

opposite ends of the harbor are two lighthouses that sound their horns every 30 seconds. Their short horn blasts are synchronous. When the rowboat is exactly halfway between the lighthouses, the man will hear both horn blasts simultaneously. If, however, his boat moves nearer to one lighthouse than the other, he will hear the closer horn first, because it takes the sound from the more distant horn a little longer to reach the man's ear. If the man knows the speed of sound and is nimble at math, he should be able to convert the time lag of the second horn to a corresponding distance. Then, having calculated how much closer he is to one lighthouse than the other, he can plot his line of position on a chart. A third lighthouse sounding its horn would enable the man to develop two more lines of position, and the point of their crossing with the first would give him a position fix. (Just one more LOP is all he really needs for this.)

Loran works on the same principle; the receiver measures the time difference (TD) between signals sent from a master station and secondary (or "slave") stations. In this case, however, the time differences are computed in millionths of a second (microseconds). Since radio waves travel near the speed of light, a microsecond of time difference is very meaningful to a loran readout. The TD between the master signal and the signal from secondary station X gives one line of position. The TD between the signals from master station and secondary station Y gives one more. (Loran receivers do not measure the TDs between two secondary station signals.) The crossing point of the two LOPs is the resultant position fix. Additional secondary stations providing additional lines of position can further increase the reliability of the fix, and most master stations are associated with three or four secondaries. The stations thus associated constitute a chain, each chain blanketing its designated area of coverage with short, timed pulses. These radio waves have a frequency near 100 kHz, giving them stable propagation characteristics.

A straight line drawn through a master station and one of its secondaries represents the "baseline" for the two stations. If your course were perpendicular to a baseline and crossed it midway between

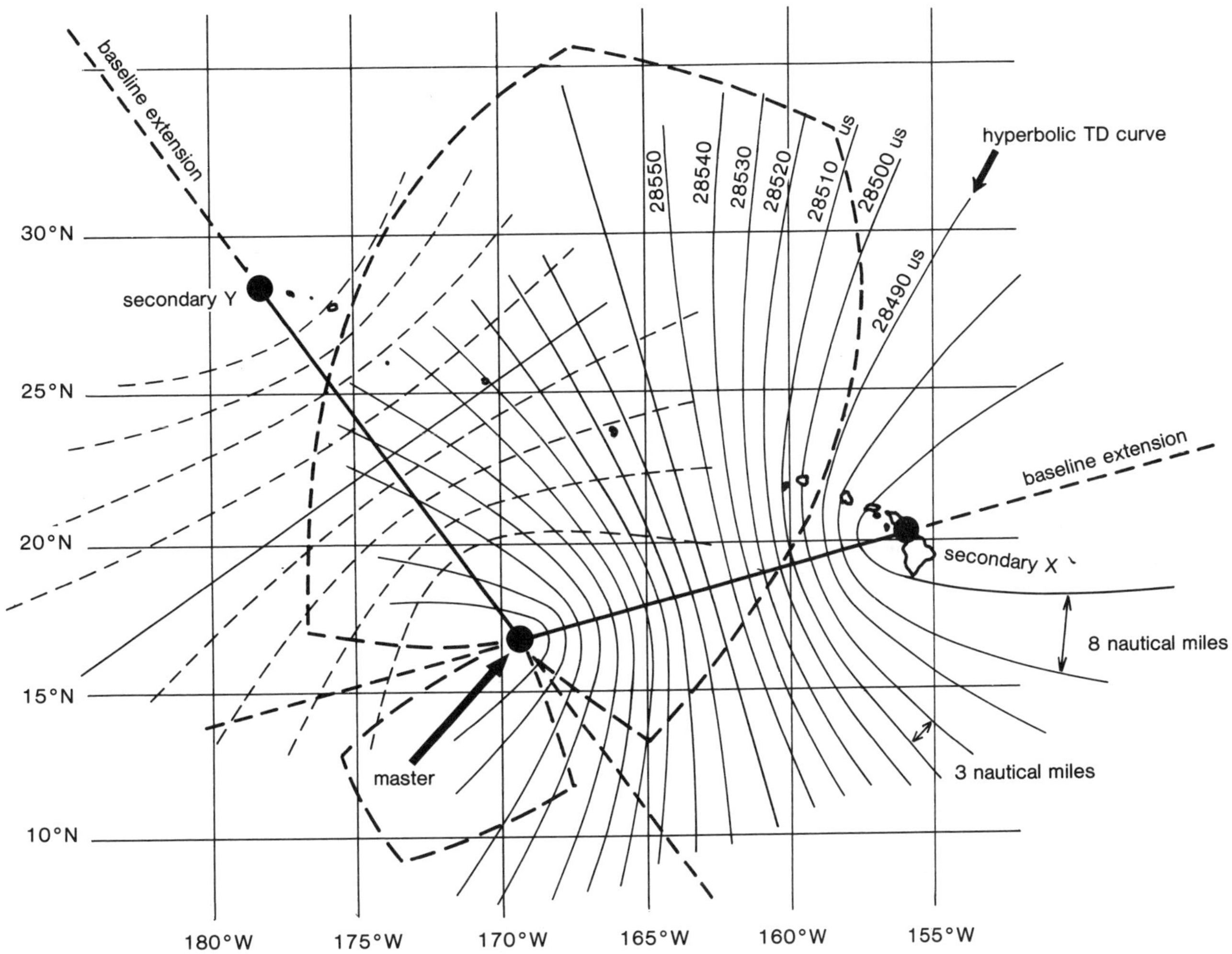

The Central Pacific loran chain, based in the Hawaiian Islands. The nominal coverage area is outlined by the heavy dashed lines. All loran chains have areas of better and poorer coverage with regard to crossing angles of the signals from the master and secondary stations. The gradient or distance between charted TDs also varies from one location to another within the coverage area, and this has a direct bearing on the accuracy of position fixes based on TDs. To make a plot using TDs requires practice and skill and is more time-consuming than plotting latitude-and-longitude positions.

stations, the time difference for transmissions from the master station and that secondary would be zero at all points along your course, provided our foghorn analogy is strictly true. Actually it isn't, quite: There would be signal interference if the master and all secondary stations transmitted at precisely the same times; therefore the pulse patterns transmitted from the stations in a chain are offset very slightly one from another. The principle is the same, however.

The master station transmits first, followed immediately by the series of secondary station transmissions from the chain. The time from one master transmission to the next is the "group repetition interval," or GRI for short. Every loran chain has its own GRI "radioprint" (like a fingerprint), and this lets a loran receiver identify the chain to which it is tuned. The desired GRI is either keyboard entered or selected or recalled from memory by the set and automatically tuned in.

Position fixing by means of loran is known as "hyperbolic navigation" because the plotted lines representing given time differences between a master and one of its secondary stations are hyperbolic rather than straight (except for the line of points equidistant between the two stations). Unless you are within 20 nautical miles of one of the transmitting stations you are using, however, segments of these hyperbolas can be treated as straight lines for plotting purposes without introducing appreciable error.

Based on the fairly constant velocity of loran signals over water and the application of theoretical or measured correction factors where the signals must pass over land, loran time differences have been printed on nautical charts. These charted TDs are a good approximation in most cases, except where local anomalies were unmeasured at the time a chart was prepared. Although the accuracy of charted TDs can be expected to continue to improve as time goes by, you should be prepared for occasional discrepancies between what your loran receiver reads out and what your chart indicates for the TD coordinates of a particular position, discrepancies of one-quarter to as much as two nautical miles. Use the charted time differences only as an initial reference, revising select coordinates as you get to know and trust your own loran readouts for local waters. You'll want to doublecheck and verify them from time to time when the weather is fair (using the positions of charted buoys, for example), but the loran coordinates your receiver gives for particular locations in local waters should be highly repeatable.

Many factors can influence the accuracy of loran position fixes. For one thing, signals may propagate skyward, bounce off the ionosphere, and return as skywaves. Horizontally propagated groundwaves yield much better accuracy, and most loran receivers will precisely time the incoming groundwave pulses, automatically ruling out interference from a slightly delayed (though possibly stronger) skywave.

Whenever possible, secondary stations should be chosen to provide TD lines with large crossing angles, the nearer to 90 degrees the better. Fixes from such TDs will certainly be more accurate and reliable than those from TDs that intersect at less than 30 degrees. Some loran receivers will automatically select appropriate secondary stations.

The U.S. government has positioned loran stations to maximize crossing angles in the coastal confluence zone, but the angles may be too small for accurate position fixes in popular cruising areas such as parts of the Caribbean, the Southeast and Gulf coasts of the United States, and the waters off Baja California. Nevertheless, the repeatability of TD fixes made in these areas should still be quite good.

Another factor in the selection of secondary stations is the "gradient" of the TD lines from that station in your locale. The tighter the gradient (the charted spacing) between lines, the more accurate a position fix. A TD interval of 10 microseconds, for example, might be equivalent to one nautical mile or nine depending on location within the master-secondary coverage area.

Finally, if you are in line geographically with a master and one of its secondaries, but not between them (that is, if you are on one of the two "baseline extensions"), position finding using that secondary will be inadvisable or even impossible. A loran receiver that automatically selects appropriate secondaries would reject such a one.

Local weather conditions may also affect loran signals slightly. For this reason some loran chains feature computer-coordinated checking receivers that monitor the signals from the master and secondary stations. These reference monitors are located near important harbors where vessel traffic will be heaviest. Some loran chains will not only self-adjust their timing sequence, but will also notify your receiver when a secondary station may be experiencing a technical difficulty. This may register on the receiver as a blinking digit or a flashing light that flags the problem secondary. This "blink" alarm will be activated, warning you to doublecheck current readings with readings taken previously, if a loran transmitter gets out of tolerance by as little as 0.2 microsecond.

Probably the biggest culprit in occluding weak loran signals is interfering noise sources, either from the boat or from one of the many powerful military stations that transmit on either side of the loran frequency band. Luckily, in the latter instance, instead of getting a slightly incorrect reading the navigator gets no reading at all. Most receivers have built-in notch filters to help eliminate this adjacent-channel interference.

Loran signals are easily masked by onboard noise sources, and indeed can actually be "wiped out" by ignition interference, electrical interference, and sometimes by noise generated on boats 50 feet away. Fluorescent lights almost always in-

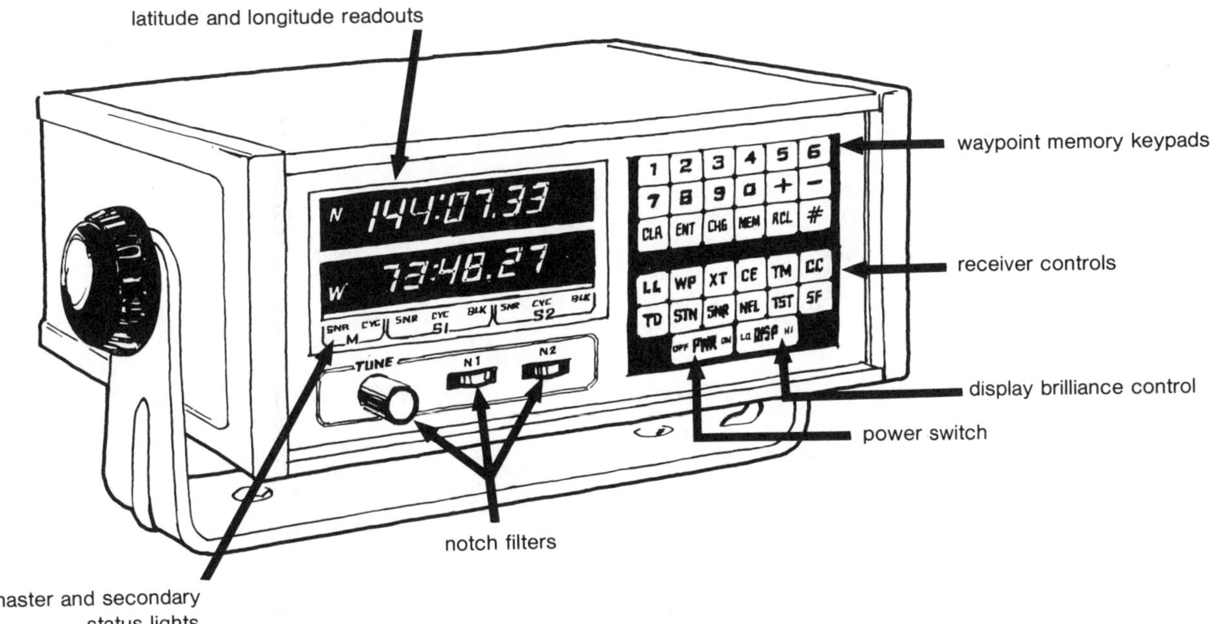

Two elements are common to all loran receivers: a display panel for either TDs or lat/lon readouts and a group of keyboard buttons to activate the various functions of the particular model. On the backs of the units are the connectors for ground, antenna, power, and (on some makes) interfaces with other electronic equipment such as autopilots, plotters, and recorders. Of great importance to the navigator is an operator's manual detailing the procedures that must be followed to utilize the many capabilities of the unit. The price range (with antenna) is \$450 to \$2,500, and an average size is 9 inches wide, 12 inches front to back, and 3 inches high, although there is little uniformity in size or styling.

terfere with Loran reception. You can actually hear the noise generated by a fluorescent light with a handheld AM radio: Tune the radio between local stations, turn on the fluorescent lights, and listen for the noise. There is very little filtering that can be added to fluorescent lights to attenuate this interference from their electronic chopper circuits. It has been such a big problem that some fluorescent light manufacturers, such as the Guest Corporation, now produce assemblies with interference rejection properties to keep them "quiet" for loran reception.

Another source of loran interference is television and video monitor displays, including raster-scan radar and video fishfinders. In these units, the horizontal oscillator sweep frequency multiplies harmonically into loran receiver range, and the resultant signal can be strong enough to wipe out even the strongest loran signals. You can test for this interference by turning video equipment on and off while watching the loran signal-to-noise ratio. If a loran set goes crazy when you turn on a specific piece of video equipment, corrective action is needed. Good grounding of both the loran receiver and the video equipment will help eliminate this problem.

The majority of loran receivers in use today can, on demand, convert the TDs to corresponding latitude and longitude readouts, which many mariners find much easier and more familiar to use. The latitude/longitude readout is essential in areas, such as at the fringes of loran coverage, where the TD curves may not be printed on charts. If the conversion is made solely on the basis of the propagation speed of loran signals over water, however, there will be losses of accuracy due to the landmass effect mentioned earlier in connection with the accuracy of charted TDs. Signals that travel over land are delayed slightly relative to those that travel exclusively across water; to compensate for this, a cor-

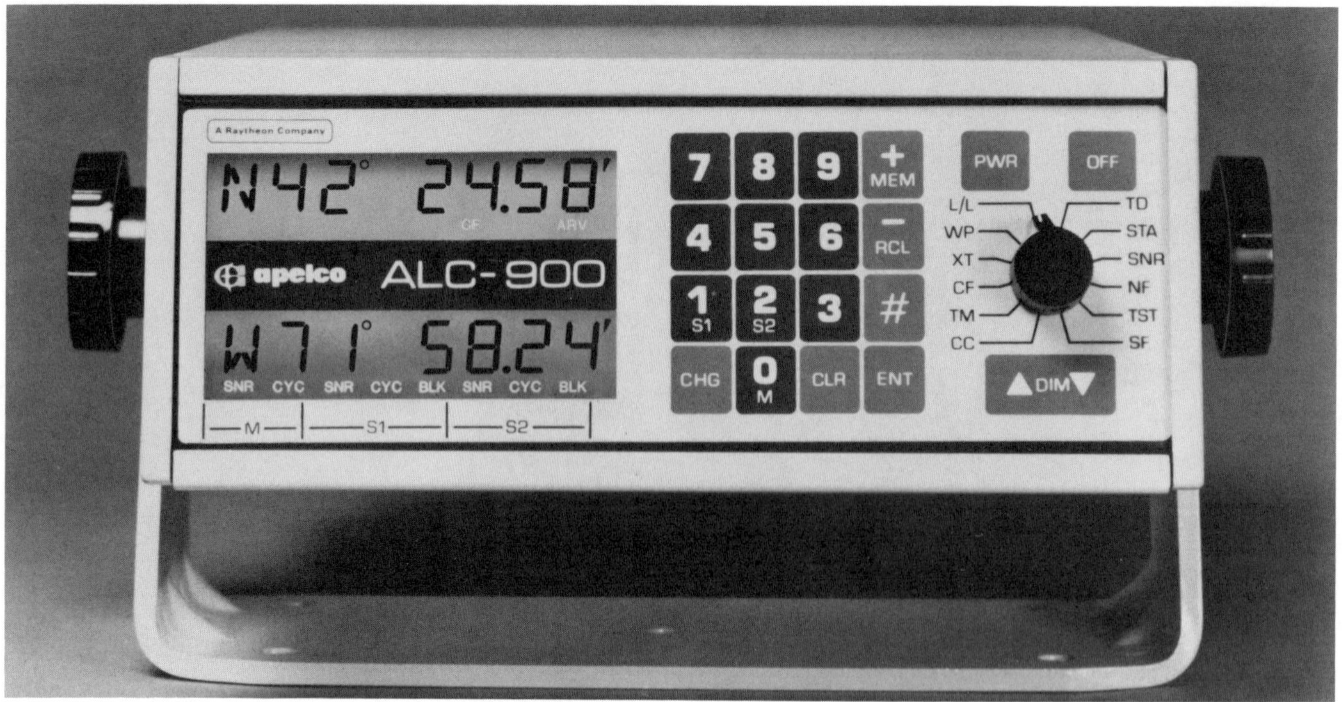

A loran receiver.

rection called the "additional secondary phase factor" (ASF) can be applied to the TDs. Some more expensive loran sets have factory-programmed ASF corrections; others require the user to keyboard enter the appropriate correction for the locale. There is also a seasonal influence when loran signals pass over land. In the spring, as trees leaf out, loran signals are delayed more than they are in winter.

What's Available (And Buying Tips)

More than 30 different types of loran sets are available today, the majority being imported, and all are good performers. Suggested list prices in the midrange are falling below $1,500. Mail-order houses are doing an excellent job of providing the most product for the least amount of money, but buying from a catalog gives you no chance to get the feel of the unit before you purchase it.

A loran set with the ability to convert TDs to a latitude/longitude readout is desirable in waters at the fringes of loran coverage or where the TD lines are not well charted. In Baja California and in the Bahamas, for example, the hyperbolic TD lines intersect at impossibly oblique angles, and it is very tough to interpolate and transfer the already-imprecise TD positions from one local chart to another. In areas like these, the latitude/longitude readout, while not, perhaps, highly accurate, is much easier to use and will be the only alternative when TDs are not charted. Unlike the accuracy, the repeatability of position fixes will usually be high in fringe areas.

As mentioned earlier, some loran sets apply ASF corrections automatically when converting TDs to a latitude/longitude readout, while others require the user to punch in the necessary corrections. The choice is a matter of personal preference, since the convenience of automatic correcting must be weighed against its expense (for the large amount of microprocessor memory necessary to store a sufficient number of local ASF factors) and the slight (no more than a few hundred yards) potential loss of accuracy it entails (because memory-stored ASF corrections must be averaged over fairly large geographic regions). ASF corrections to be entered manually can be taken from charts or from the *Loran-C Correction Tables,* published by the U.S. Defense Mapping Agency and available from chart houses. In areas not covered by these publications, mariners can develop their own ASF corrections by comparing latitude/longitude readouts with the charted positions of precisely located navigation buoys. A single ASF correction is likely to be accurate within a 20-mile-diameter circle, but there are local exceptions. Experience is the best guide.

All loran units are equipped with notch filters to screen out unwanted noise from frequencies near the loran 100 kHz band. A good set will have three or more, and in most receivers at least two of these are calibrated by the dealer to screen out known noise sources in the buyer's local waters. Be forewarned that a discounter selling by mail order, if it is not a factory-authorized dealer, may lack the facilities to "tune" a receiver for your locale.

Some receivers have, in addition to preset notch filters, other internal filters that automatically self-adjust to weed out unwanted signals on surrounding frequencies. Alternatively, there may be manually adjustable filters—either external or internal—that can be retuned for new cruising areas and new environmental or on-board noise sources. These may be desirable near large metropolitan areas or to combat high shipboard noise levels from engine ignition systems, fluorescent lights, bilge pumps, tachometers, etc.

The type of display is another variable among loran sets. Compared to gas discharge tubes and light-emitting diodes, liquid crystal displays are a little easier to see and getting more so. At night they are backlit for easy viewing, and they consume almost no power at all. The degree of resolution of the display also varies, many units displaying TDs to the nearest tenth of a microsecond, a few displaying them to the nearest hundredth of a microsecond. The latter might be considered overkill, implying more precision than even loran can habitually offer, since 0.01 microsecond is equivalent to less than 10 feet on most TD gradients.

Waterproof or weatherproof models are available for open-cockpit use. Some loran units are built in submersible cases and are ideal for open boats or small runabouts. Their pressure-sensitive keyboards keep moisture out, are made of a newly developed membrane that is not vulnerable to in-

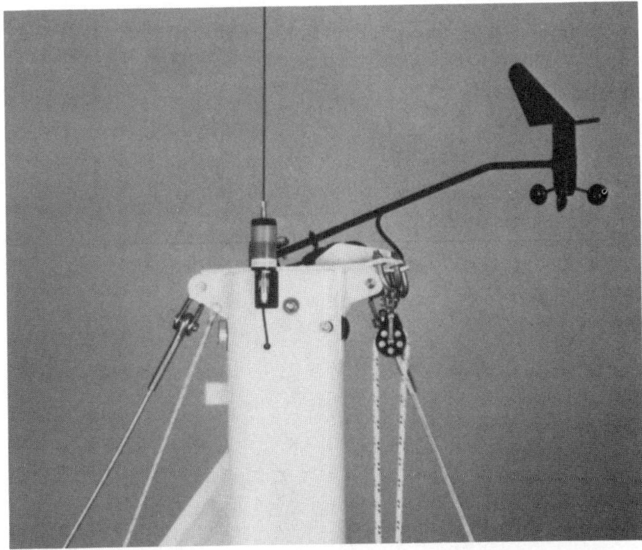

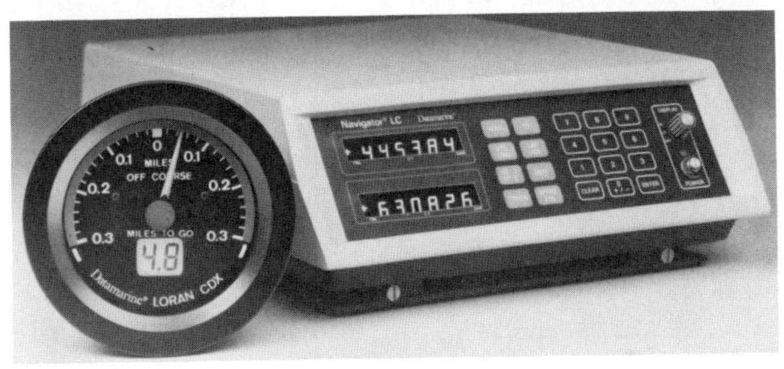

The Datamarine loran system package, including mast-mounted whip and cockpit readout of cross-track error and distance to run.

tense sunlight, and will probably last for years if kept clean and saltfree. These units may be portable, with a rechargeable battery pack.

It is also possible to purchase a unit with a waterproof remote-control "head," enabling operation of the receiver from the cockpit while the unit itself stays dry below.

Most if not all loran receivers are sensitive enough to receive the needed signals and obtain a position fix anywhere in the coastal confluence zone, and the repeatability of fixes is universally good. Blue-water voyagers and those intending to cruise extensively in the fringe areas of loran coverage should be aware, however, that there are differences among receivers in the ability to pick up a weak signal, separate it from unwanted noise, and use it as the basis for a fix. For those comparing manufacturers' specs, a "dynamic range" of 120 decibels (the norm is 110) and an "acquisition signal-to-noise ratio" of −18 to −20 (the usual is −10 to −15) might indicate a heightened ability in this regard (and a higher price tag), but probably there is no substitute for the advice of experienced users.

For most mariners, the matter of preference is decided by the front-panel controls—how easy is it to make the unit do its thing properly? Imagine that you are turning on a receiver and preparing to use it. Some units will require the operator to select and keyboard enter the GRI of the chain he wishes to use, while others will either use the last chain entered or automatically choose the best one for the locale. In the latter case the choice is made on the basis of signal strength levels, signal-to-noise ratios, crossing angles of the TDs, TD gradients, and the proximities of baseline extensions. These units will apply the same criteria in selecting the most advantageously placed secondaries within the chain. In essence, your thinking is done by the loran's microprocessor, a feature that is best appreciated when the navigator is in unfamiliar waters.

Once the unit is turned on, it may require 5 minutes to settle down and start giving consistent readouts. Some need less than 2 minutes, but it's usually best to allow at least five. In the interim, a more sophisticated unit will go through a self-testing mode while you passively wait. It will doublecheck every integrated circuit, perhaps telling you through its alphanumeric display in fairly plain English, "Testing receiver." If it has automatic notch filters it will examine the frequencies around the loran band and adjust the filters as necessary, perhaps telling you, "Testing and adjusting filters."

When the unit is ready, it will let you know with a beep or a flashing light. Usually the time differences appear first. Assuming the receiver has a latitude/longitude conversion, how easy is it to call up this alternative display? Some units do it with the push of a single button.

Most loran receivers today do much more than position find. Just about all of them permit the programming of "waypoint" coordinates, a waypoint being a preselected location at which the navigator wishes to change course or make observations, or to which he wishes to return later. It is possible to reach or return to waypoints with a precision of 50 feet or better. Some sets can memorize more than 100 waypoints, and the number will continue to rise in the future. Let common sense prevail. While it may be handy to store a number of selected waypoints representing buoys or landfalls in your local waters, few mariners will use more than five or six waypoints in a single passage. The receiver may also give the bearing (possibly adjusted for local magnetic variation) and distance to the next or the last-visited waypoint when it is "asked" for this information by the pressing of one or two buttons. It may also tell you how fast you are going and how long it will take to get there, and it may beep when you approach or reach the waypoint. Some units will automatically inform you when and how far wind or currents are setting you off track for the next waypoint, and they will suggest a new course to steer; you can set a "cross-track" alarm to go off when you stray farther than 0.05 to 9.99 nautical miles off the desired course.

Other features are available—some useful, some more or less frivolous. These include clocks, alarm clocks, interval timers, and stopwatches. Some sets can be reprogrammed for waypoints at home, having backup batteries or memories that are not erased when the juice is turned off. Many units have a panic button that, when pushed, will store the precise current position in memory, enabling you to return to the spot where someone or something has been lost overboard.

With so many capabilities, sophisticated loran units frequently are not exploited to their fullest potential by their users. Some receivers are more "user friendly" than others, perhaps prompting the operator by means of messages on the screen to take the next step in entering waypoint information or requests through the keyboard. Manufacturers are sensitive to the fact that they have created something so sophisticated it may be scaring mariners off, and they are trying to humanize the equipment. One unit will flash "Baloney" if you try to enter a waypoint that's in the middle of a landmass, or "AYE AYE" in response to a keyboard-entered request.

Autopilot Interface

Many loran units will tie into an autopilot, and the National Marine Electronics Association has agreed on a standard 0180 autopilot output to make the interfacing easier. It is best to consult the services of a marine electronics specialist for this, but if both loran receiver and autopilot were manufactured after 1981, they probably have the appropriate inputs and outputs. Commercial fishermen can find this an invaluable tool.

To supplement the discussion above, here is a list of other features you may wish to consider when shopping for a loran receiver:

- Ability to reject skywaves in favor of ground-waves, if present
- Skywave warning indicator
- Loran chain false indicator
- Loran chain error indicator
- "Off the air" secondary station identifier
- Signal-to-noise ratio figures
- Low signal alert
- Permanent memory that is erasable only on special command
- Automatic sequencing of several displays
- Cross-chain interference rejection
- Convenient mounting brackets supplied
- Low power consumption
- Numerous service stations in case of repair throughout the world, which is a function of name brand
- Name-brand equipment with proven track record
- And of course, competitive pricing

Installation

A loran set is a very sensitive radio receiver without a loudspeaker; it can be installed by the boat owner provided proper care is taken. The necessary small current draw from a 12-volt source is easily tapped from the electrical panel. It is best to use an independent 12-volt line that goes directly to a circuit breaker, the breaker, in turn, having a direct connection to the battery. The 12-volt wiring doesn't necessarily need to be big—a pair of No. 12 marine duplex wires will work just fine. Wiring directly through a breaker (for safety) to the ship's battery system eliminates noise sources that might travel up a power lead from another connection point. A great deal of noise, for example, is generated in the chopper circuit of fluorescent light controls, and this noise can easily jump over to a loran power lead and cover up a weak signal. If tapped directly,

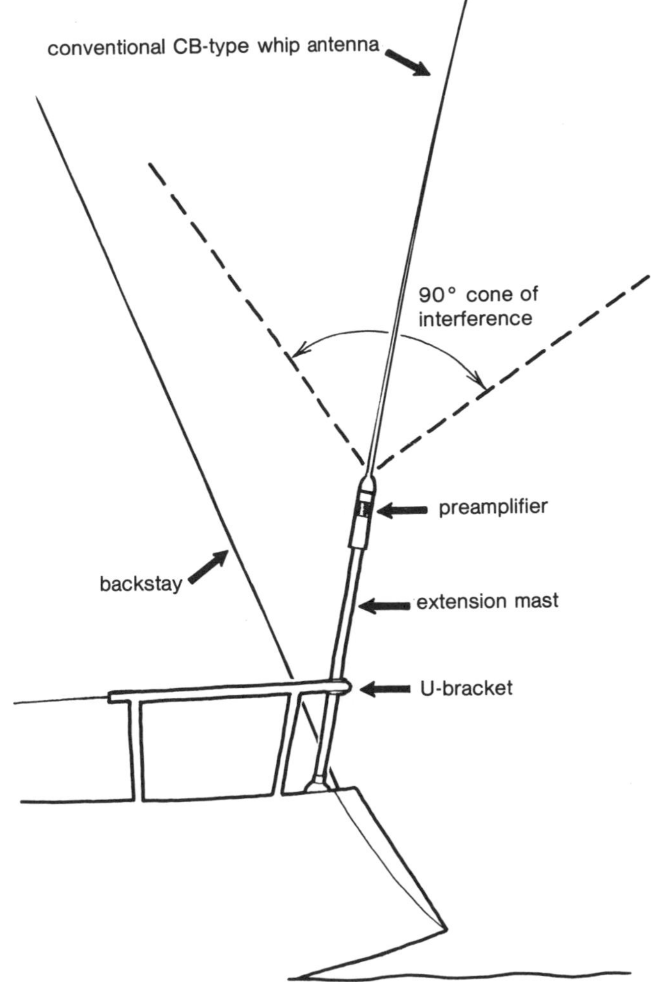

A transom-mounted loran antenna on a sailboat. The height of the antenna is not critical, but the angle should be more than 45 degrees above horizontal. Regular CB-type antennas are satisfactory and relatively inexpensive. The coaxial cable from the preamplifier is fed through the extension mast to the receiver at the navigator's station.

the ship's battery will act as a giant filter to help "smooth out" random noise pulses found in other parts of the marine electrical system.

Since the loran receiver draws only about one-half to one amp, there is little worry that it's going to pull down your starting or auxiliary battery if you happen to leave your set on over the weekend. In fact, the current consumption is so low that you *could* power the set with a dry cell battery.

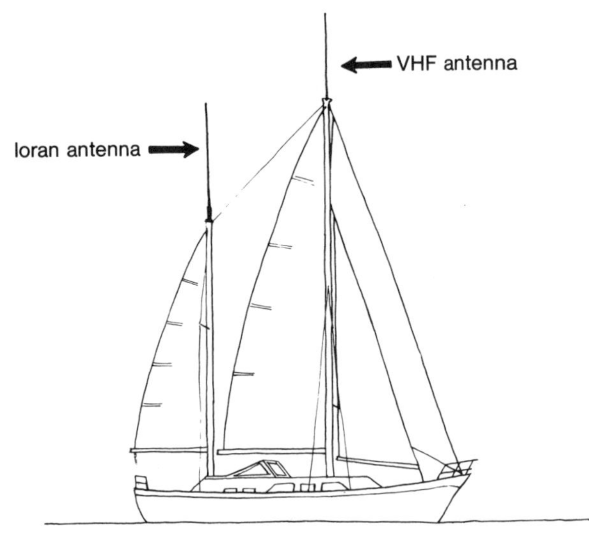

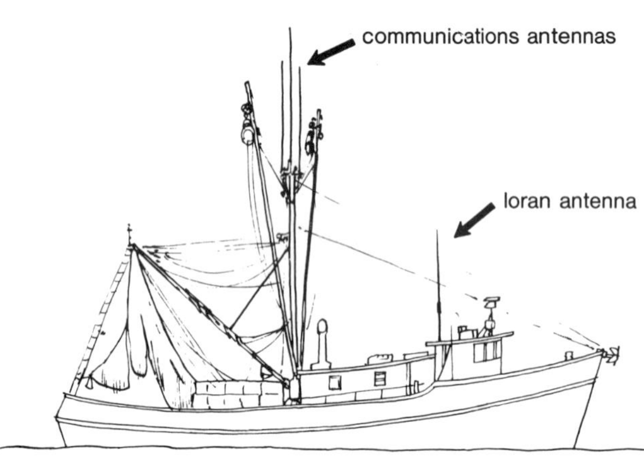

When power or sailboats have antennas for VHF or SSB equipment, the loran antenna should be mounted as far from them as is practical. Since any antenna is subject to damage, a spare whip should be aboard.

Like any radio receiver, a loran set needs a good antenna and a ground system, the latter to enhance weak signal reception and to help the receiver sort out weak groundwave signals from a high noise level. A good ground is also necessary to prevent dry-weather static discharges from erasing any of a set's waypoint memory. To get the most out of a receiver, a groundplane of 60 to 100 square feet is probably necessary. Components of this groundplane might include keelbolts, the engine block, through hulls, etc., as well as copper screen and copper foil when needed. Copper foil should be run from the receiver chassis and the antenna base to the groundplane. Methods of achieving a proper groundplane are discussed in more detail in the chapter on single sideband radio; while the procedures may be tedious and time-consuming, the standards of quality need not be as high for the receive-only loran unit as they are for a single sideband transceiver. An owner installation is therefore certainly feasible. (If you do intend to have a professional do the installation, buy the receiver from a marine electronics dealer and include the price of the installation in the purchase package. It will be less expensive that way than on a time-and-materials basis.)

It makes sense to use the antenna that a loran manufacturer recommends, particularly since most receivers are shipped with a matching antenna system. Most loran antennas, however, are nothing more than 9-foot white fiberglass CB whip aerials. The whip should be mounted out in the open and as far away from noise sources as possible. The heart of the antenna system is an encapsulated pre-amplifier, which the whip screws into. This pre-amp magnifies the incoming signal level and creates a narrow pass band for loran frequencies to enter. You may wonder how this low-noise pre-amplifier works without a DC voltage cable to provide power to it; current is indeed supplied to it, but that current uses the same coaxial cable that carries the incoming signal from pre-amp to receiver. One coax cable doing two jobs simplifies the antenna installation. If you do not know how to solder coax connectors, use the crimped, no-solder connectors discussed in the chapter on VHF radio. If you should ever break the whip, any CB radio-type whip will do quite nicely on the pre-amplifier mount. Most loran whips have a $3/8$-by-24-inch thread, common among CB antennas.

Height is not important for a loran antenna; anywhere in the clear is fine. You can even rake the antenna back at an angle of up to 45 degrees without significant signal loss. Just keep it away from fluorescent lights, bait tank motors, refrigeration systems—anything that generates electrical noise.

Troubleshooting

If your set does not function properly when the engine is running, noise-suppression steps are probably in order. Filters to eliminate ignition interference are available from marine electronics specialists, as are filters specific to just about any other noise source. Diagnose intermittent reception by having another person watch the display while you turn on and off possible noise sources.

Firm connections to antenna and ground circuit are essential. Any loose or corroded connections should be cleaned, tightened, resoldered, or replaced as necessary.

Nearby military transmitters can mask loran signal reception. If your receiver doesn't work in one locale but does work everywhere else, this is probably the problem. Factory or dealer adjustment of internal notch filters may be necessary. You may be able to avoid this by purchasing the unit from a dealer who is familiar with the interference sources in your local waters.

10

Noise Suppression

by Gordon West

Electrical interference noise is the nemesis of modern marine electronic gear. Whether noise is generated by a boat's engine or originates from another onboard source or an inadequate installation, any marine electronics dealer that installs gear has an obligation to the boat owner to trace the source of the problem. Common practice when the cost of a major marine electronics package (including installation) is quoted by a dealer is to leave ample room for an extra day onboard knocking out problems of electrical noise. When you install your own gear, chronic interference is your problem, not the gear manufacturer's or the dealer's. This chapter should help you locate and eliminate the more common sources of noise.

Electrical noise is not hard to spot on communications receivers, but on specialized pieces of marine navigational equipment, noise can sneak in without your knowing it. To begin this discussion, let's look at the common symptoms that accompany gasoline engine–related noise in marine installations:

- Popping and clatter on VHF set with engine turned on; completely quiet with engine off.
- Single sideband signals disappear into popping and clatter when engine is turned on.

- Your sophisticated audio system "sings" and buzzes, with popping and clatter on weaker AM and FM signals.
- Speckles and rolling bars on the TV set when the engine is turned on.
- Loran signals disappear, and the signal-to-noise ratio indicator on the loran set indicates poor reception when engine is on; with the engine off, loran works fine.
- Low-level satellite passes are not picked up by the SatNav receiver with the ignition system on.
- OMNI/VOR reception is impossible just a few miles from the transmitting station.
- Low-frequency Omega, loran, and RDF reception is impossible with the engine running.
- Speckles appear and weak targets disappear on radar scopes.

The recurring theme is that as soon as you turn on the engine the noise comes in, and there is little question that the noise is generated by the engine and its ignition system. The first reaction is to blame the sophisticated electronics for being incapable of filtering out the noise; the fault is not with the electronics, however, but rather with the interfering source. As radio receivers are made ever more sensitive to weak signals, their sensitivity to noise increases in direct proportion. These electronics are not *supposed* to filter out noise, even though some offer noise-blanking and noise-elimination circuits. Such circuits only help the effort; the real steps must be taken at the source. According to Jack Honey, a marine electronics and amateur radio noise elimination specialist, "Tracking down and eliminating electrical noise leaks is just like plugging leaks in the bottom of a boat. The job isn't done until every leak has been stopped." Honey's company offers a complete line of EMI noise elimination filters and instructional sheets. (Marine Technology Corporation, 2722 Temple Avenue, Long Beach, California 90806; 213/595-6521.)

Finding The Leaks

The hardest part of quenching electrical noise is simply locating the point source of the noise being radiated or inductively coupled on the electrical wiring. Once the source is identified, filters and shielding will quickly and effectively cure the problem.

One method of searching is to listen to the characteristic sound of the interference on a communications receiver—or better yet, on an inexpensive pocket AM radio with an earphone. Most noise is broadbanded, with the greatest intensity on frequencies that an AM radio will receive. In order to pinpoint the noise, you may need to wrap the radio in aluminum foil except for a small hole near the built-in loop stick antenna. This shielding will give you highly directional reception, enabling you to move the tiny radio around in the engine compartment to good advantage as you attempt to locate the source.

Another way to find a hot spot of noise is to take apart a defunct cassette player and mount the playback heads on a 3-foot wooden or plastic probe. Using shielded wires, connect the heads to a simple audio amplifier with a high-impedance input. Now move the head around until you pick up the noise you are experiencing.

Following are some of the noises most commonly found aboard boats, motorhomes, cars, and airplanes:

Type of Noise	Possible Interference Source
Distinct pops at idle, increasing to roar at high engine speeds	Gas engine ignition system
Popping sound that increases in intensity as engine speed increases	Electronic tachometer
Musical whine, increasing in pitch to a whistle at high engine speeds	Alternator or generator
Intermittent frying noise	Voltage regulator
Whine or buzz from certain accessories, even though the engine is turned off	Accessory motors or power converters
Crackling noise when certain lights are turned on	Fluorescent lights
Constant-pitch low-frequency noise when entertainment electronics are turned on	Television receiver
Persistent noise on a receiver even when its antenna is disconnected	Conducted noise on the power line or from switcher power supply
Grinding noise when in gear	Intermittent electrical grounding of shaft

Finding, Filtering, And Isolating Ignition Interference

"Pigtail" capacitors have traditionally been used for filtering ignition noise, and they work reasonably well at low and medium frequencies. They are completely ineffective at VHF frequencies, however, because of the high inductance of the long lead. Don't expect these capacitors to be an all-out cure for ignition noise.

Let's start with the engine. There are plenty of "switches" in an engine's electrical system that will generate noise. The breaker points in a gasoline engine make and break the circuit from the battery through the coil to ground perhaps 200 times per second. Without good noise filter suppression techniques, the interrupted current and its associated radio frequency impulse noise will flow in the wiring from the coil back up to the ignition key, and from there back to the battery and the ground. This makes a very large and effective loop antenna to radiate noise throughout the vessel. If these wires are cabled in with other wires, RF impulses will be coupled into the companion wires as well.

Another concern is the small arcs in the high-tension part of the distributor and in the spark plugs themselves. These arcs generate radio frequency noise, which can be radiated by the plug wiring and is also coupled through the coil into the primary circuit wiring, where it joins the RF energy produced at the points.

Other problems include alternator diodes, which cause a sudden change in current when they go from forward conduction to cutoff. The older, DC generators and motors have brushes in their moving commutator segments that switch current in both the internal armature winding and associated external circuits. The voltage regulator has a vibrator that chops the field circuit thousands of times per second when it's regulating, with resultant RF noise flowing throughout the wiring harness.

Replace old spark plug wiring with new resistant-type wiring. One such wiring construction features a fiberglass core impregnated with a conductive powder, and it works well in knocking down the popping noise from gas engines. This type of wiring must be changed every two years, however, because of its susceptibility to failure in a moist environment. Belden Corporation markets a spark plug wire under the NAPA label that uses a conductive neoprene tube and seems to stand up better aboard boats. Contrary to persistent rumor, the use of resistance wire to the plugs does not appear to involve any sacrifice whatsoever in engine performance if the balance of the ignition system is in good shape. Jury rigging plug wires using coax cable with a grounded outside braid will only foul up the ignition system timing.

Use of both resistance wire and *resistor plugs* is more effective than the use of either one alone. Be sure and check your owner's manual for the proper plug type and gap for your engine.

A low-pass or feed-through capacitance filter at the coil in series with the connection to the ignition key and any other connection made to the battery side of the coil is an absolute must. The usual

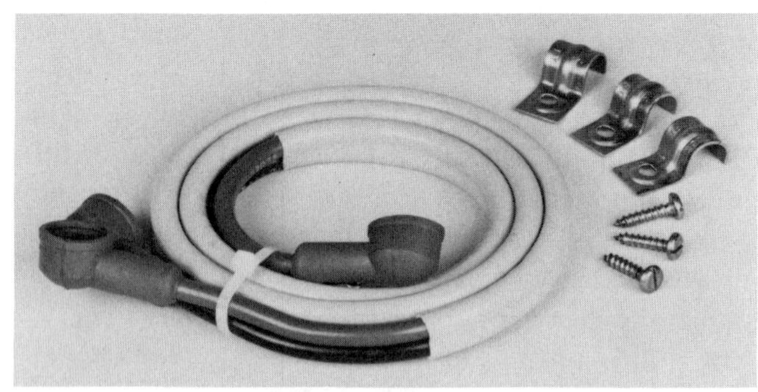

Resistance wire with resistor plugs

choice is a low-pass, PI-type LC filter rated at 5 amps, with good operative characteristics from 0.1 MHz through 300 MHz. The filter should be mounted solidly to the engine or the coil bracket, as close as is convenient to the battery terminal of the coil. Keep all filter wires as short as possible, and make sure that the bracket securing the filter is well DC grounded.

Electronic tachometers that pick up 400-volt impulses on the coil will usually radiate noise along the wire to the tach readout unless the entire wiring harness is encapsulated in a ground sheath. Many aren't. If yours isn't, you should shield the tachometer wiring as well as install a tachometer filter at the coil in series with the lead to the tach. Use a low-pass RC filter, which won't load the ignition circuit with additional capacitance yet passes enough signal to operate the tachometer.

Again, you can spot these noise sources by using a transistorized AM radio with an earphone, or a cassette tape pickup head wired into an audio amplifier.

Alternator noise, which is easily identified by the musical whine it generates, comes from the typical 3-phase, 14-pole, AC alternator with a DC field and a built-in 6-diode, fullwave bridge rectifier circuit. The level of whine varies with output current, and thus is controlled by the regulator and the amount of connected load. It may fade to almost nothing as the regulator cuts the charge rate to a fully recharged battery back to just a few hundred milliamps. A 60-amp low-pass filter in series with a heavy alternator output lead is a must; small alternator filters installed at the affected receiver will generally offer only 50 percent relief. The proper filter is a PI section, LC, low-pass filter with low

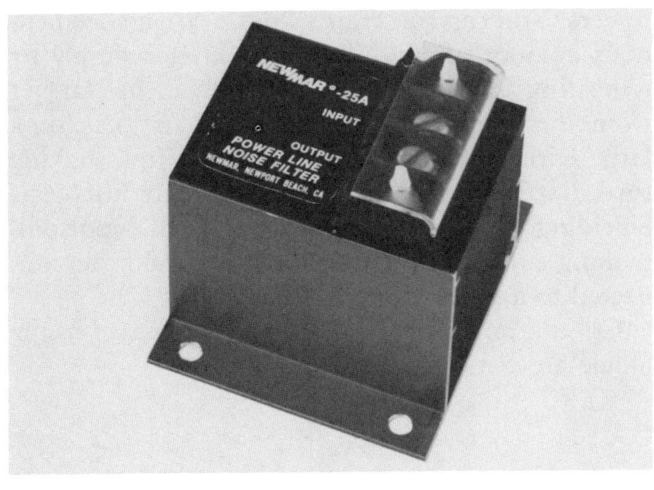

Power line noise filter

Transceiver noise filters

Radio noise filter

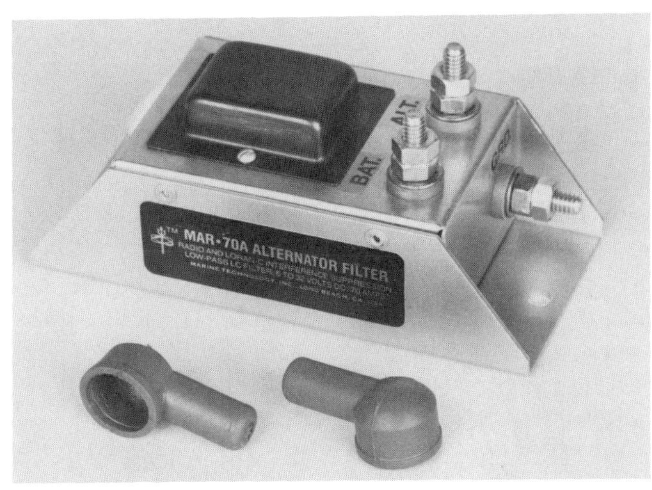

Alternator noise filter

lead inductance, rated to 60 amps continuous from 6 to 115 volts AC or DC. If you're working on a system that uses an old-fashioned generator, the same type of filter is used.

Voltage regulator noise elimination is a bit more tedious. Additional grounding and shielding of the voltage regulator case may help. This done, locate the field wire and possibly a second light-gauge wire running from the alternator to the regulator. If these are cabled in with other wiring, they should be run separately. You might also disconnect the existing wires and substitute shielded wires, making sure the shield of each wire is grounded at both ends. RG-8 coax cable is ideal for this shielding.

You can also filter the regulator wiring with 0.2 mfd, 200-volt capacitors if you keep the leads very short. Never bypass the field lead between the alternator or generator and the regulator with these capacitors, however.

Small motors that radiate noise when turned on can be bypassed with these same 0.2 mfd, 200-volt capacitors between the hot lead and the case of the motor, keeping leads very short. If the capacitors can be mounted inside the case from each brush to ground, all the better. Specific filters available from leading noise filter manufacturers actually contain three capacitors specifically designed for quieting these accessories.

In the case of television or fluorescent light interference with loran, Omega, or RDF systems, you can try relocating the receiving antenna, or, easier yet, simply turn off the TV or fluorescent lights when operating the receiver in weak signal areas. Newer fluorescent lights for motorhomes and boats now have special noise elimination circuits to keep the energy from being radiated down the line.

Conducted interference may sometimes occur if large high-current wiring bundles are located near a sensitive audio tape player or radio frequency receivers. You can sometimes knock down a tremendous amount of conducted interference simply by relocating a wiring bundle just a few inches farther from the receiver. You might also try shielding large wiring bundles with aluminum foil, making sure that the foil does not contact any hot wire. Shielding audio speaker leads is also important, making sure that both ends of the braid are connected to a good ground source.

Good grounding techniques are discussed in the Single Sideband chapter.

11

Satellite Navigation

by Gordon West

A satellite navigation receiver will display your position within 100 to 200 yards in any kind of weather—close to shore or thousands of miles out at sea. It will operate anywhere in the world. It displays your position as latitude and longitude, and there are no correction factors to apply (as there are for Loran-C). Prices have fallen below $1,500. The highly sophisticated microprocessor-based receiver is no larger than a VHF radio, and the antenna is also small. The satellites and associated ground stations upon which this navigational system depends will be maintained until at least the end of 2000, and there will be five years' warning prior to the system's termination. (Its eventual replacement will be an even more impressive satellite array already dubbed the Global Positioning System.) If you buy a SatNav receiver in 1988 you can expect to get the better part of a decade's service from it, and probably more.

The satellite navigation system, then, has a lot to recommend it. No land-based system, including Omega and loran, can compete with the accuracy of a fix taken on an overhead satellite. Satellite navigation is the wave of the future.

How It Works

The existing satellite navigation system, called the Transit system, was developed by the United States Navy in the 1960s to give Polaris submarines a quick position fix with minimum surface time. These occasional surface fixes were necessary to recalibrate the submarines' on-board Inertial Navigation Systems. (Today, submarines use *both* Omega and the Transit system.)

After the Soviets launched the Sputnik I satellite on October 4, 1957, radio operators receiving its transmissions in the United States noticed a Doppler shift as the satellite approached, passed overhead, and then receded over the horizon. As the satellite was approaching, the transmitted frequency appeared several kilohertz higher than it actually was because of the velocity of the approaching "bird." This frequency offset disappeared momentarily as the satellite passed through the local zenith, only to reappear as a lower-than-normal apparent frequency when the satellite retreated. It was possible using this Doppler shift to

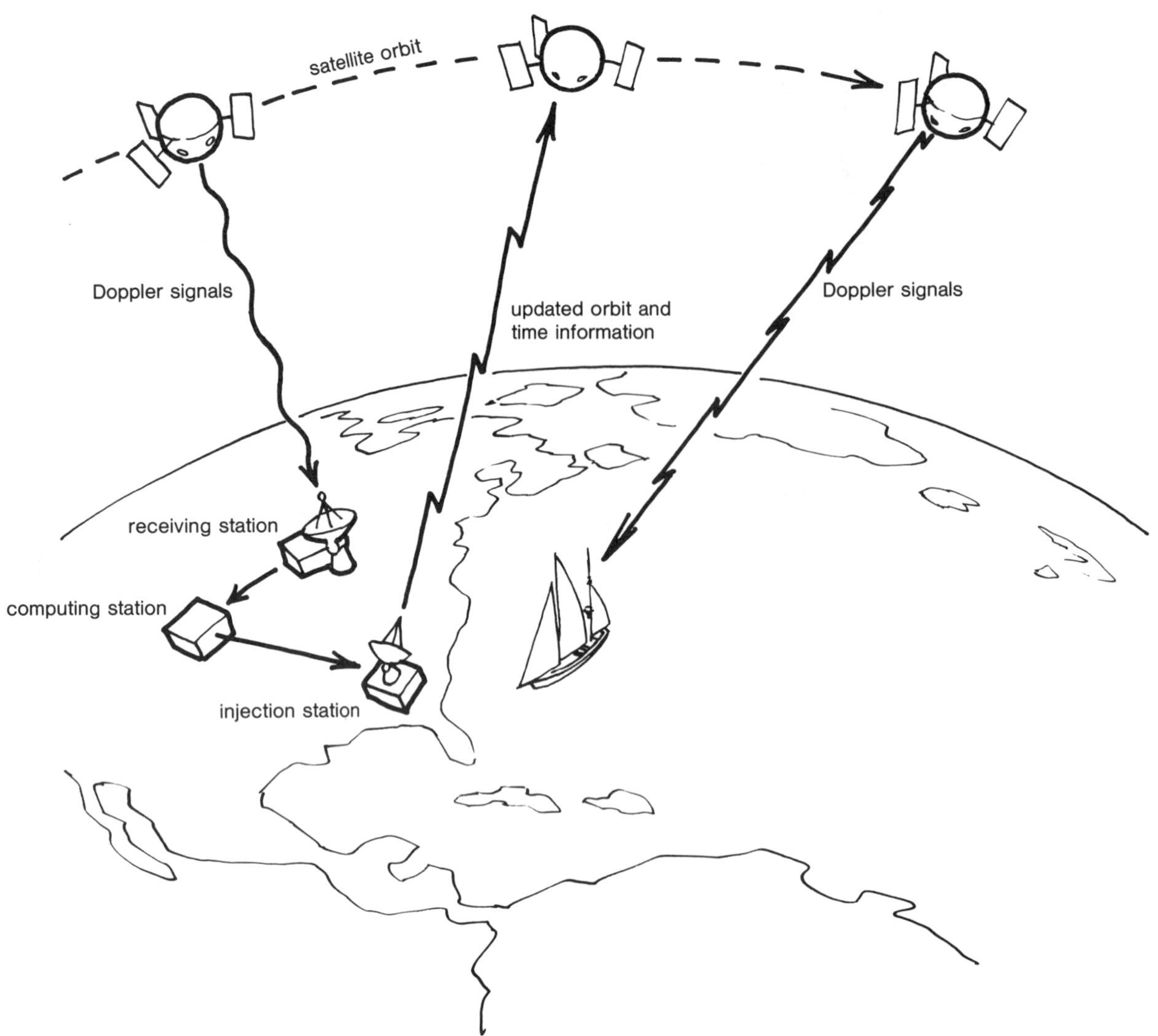

The satellites of the Transit system orbit around the north and south poles. Ground stations monitor the Doppler signals and periodically transmit to the satellites the precise and up-to-date parameters of their orbital paths. The satellites then transmit this information to vessels for position fixes anywhere in the world.

formulate algorithms by means of which upcoming orbits and future satellite passes could be predicted. These algorithms proved extremely accurate, and they could be developed from Doppler shift measurements made at a single, fixed ground station.

By logical extension, if we know precisely where a satellite is, we should be able to use Doppler shift measurements as a means of fixing our position. This is the basis of the satellite navigation system. The satellite orbits are known and predicted, and with this information a SatNav receiver can tell us almost *precisely* where we are in terms of latitude and longitude.

The Navy first allowed the use of their Transit navigation system by civilian mariners in 1967. The system consists of five polar-orbiting satellites circling the earth every 107 minutes at an altitude of about 700 miles, their paths paralleling earth's meridians. This means the satellites are whizzing along at speeds in excess of 15,000 miles per hour. The satellite orbits are very close together where they pass over the poles, reaching their greatest distance of separation where they cross the equator. The orbits are fixed, creating the effect of a cage within which the earth rotates. Thus we see a different satellite on every pass. Although the time between local passes for position fixing is two to as many as six hours in equatorial waters, it varies between 35 and 95 minutes everywhere else. You can receive an updated position fix at least every hour in most of the world's cruising areas.

The five orbiting satellites are tracked by the Navy Astronautics Group in California, with associated tracking stations in Maine, Minnesota, and Hawaii. These stations monitor the orbits and periodically update the satellites' "on-board" navigational systems with commands transmitted from ground "injection" stations. Each satellite receives a new injected message about every 12 hours, and is thus always able to communicate its position with pinpoint accuracy. This system has been phenomenally successful over the past 10 years.

Every two minutes, each satellite transmits a navigational message intended for line-of-sight shipboard and land-based receivers. This encoded digital message contains precise orbital information that a SatNav receiver uses in calculating the satellite fix. While your receiver is picking up this two-minute "station break," it is also measuring local time and gauging the Doppler shift created by the approaching and departing satellite, correcting this for the slight Doppler shift caused by your own vessel's movement. Using these data, it computes a line of position from the satellite's known location. During the approximately 15 minutes the satellite is well above the local horizon, several LOPs are thus obtained and their best intersection determined to fix the vessel's position, which is displayed as a latitude and longitude readout on the face of your satellite receiver. The receiver gives you a beep to let you know when a satellite is beginning its transmission, and another beep 10 to 15 minutes later to inform you that all of the information has been processed and your position is displayed on the LCD readout.

The satellites broadcast on radio frequencies in the VHF and UHF region using a tiny one-watt transmitter. Their huge antenna systems, which point down toward the earth, intensify outgoing signals. Internal batteries charged by solar cells keep the transmitter going 24 hours a day, and an elaborate system of counterbalancers keeps the satellites' directional antenna pointing always toward earth.

Local weather conditions, day or night ionospheric variations, sun spots, solar disturbances, and radio "blackouts"—none of these have any effect on the incoming UHF radio waves. About the only thing that can adversely affect SatNav reception is an impediment or obstruction in the line-of-sight path between your SatNav antenna and the satellite. On some passes the bird may be barely above the local horizon, and reception will be weak. In this case, you will not get pass information, and you must wait until the next satellite comes around in another hour or so.

It has been found that the satellite transmission can reach your small (about 3 feet long) satellite antenna even if someone is standing in its path. A metal object between the antenna and the orbiting satellite, however, might cause it not to "see" the incoming signal, so don't dry out a dish pan or metal fishing lures by putting them over the antenna. Try to put the antenna out of the way, but don't worry too much about a sailboat's aluminum mast—most satellite signals will filter through the rigging and around the mast and can be collected by the antenna quite nicely. Mounting the antenna too close to a powerboat's steel stack might reduce operating capabilities. The height of the antenna is not critical. Once the antenna is installed, simply keyboard-enter the antenna height into your receiver's microprocessor memory.

Between satellite fixes, the SatNav receiver displays a continuously updated dead reckoning (DR) position. This requires that course and speed information be entered into the receiver. (The receiver also needs this information in order to make accurate computations from Doppler shift measurements during satellite passes.) If you are on a steady course and speed, you can enter the information manually via the receiver's keyboard, and the receiver will give you, in addition to an updated DR position readout, the precise time that you can expect your next satellite position fix.

A far more efficient and accurate method of inputting speed and heading information for a receiver's DR calculations—especially on sailboats, for which steady speeds and courses are the exception rather than the rule—is to use a fluxgate compass and a speed sensor that are electronically coupled to the receiver. You would do well to use

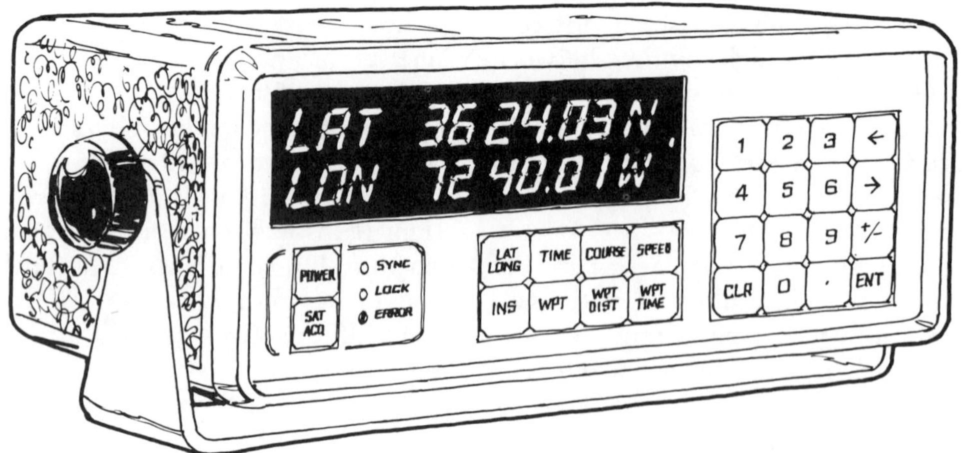

Satellite navigation receivers display the current latitude and longitude, which can be plotted on the appropriate chart. Using the keyboard, the operator can prompt the receiver to display many other types of navigational information, such as distance to a destination, course to be steered, speed made good since the last fix, total distance traveled, and even the time within one second. The price range (with antenna) is $1,000 to $5,000, with the more expensive units being capable of tying into such instruments as an autopilot, knot log, and compass. The size of a typical unit is 12 inches wide, 8½ inches front to back, and 4 inches high.

the heading and speed log systems recommended by each SatNav manufacturer. While it is possible to interface a new SatNav receiver with your present speed measuring system, the attempt may cause a lot of headaches, and it will complicate the installation process. Find out ahead of time whether the SatNav receiver you are thinking of buying will interface readily with the speed log you already have. Most satellite manufacturers produce their own fluxgate compasses that mount almost anywhere for heading information. Use the compass offered by your receiver manufacturer. Once you have these electronically coupled sensors in place, your SatNav receiver will receive continual updates on heading and speed changes.

The SatNav receiver's DR calculations are subject to the same inaccuracies—especially those caused by currents and leeway—that affect all DR plots, the difference being that with each new satellite fix the slate is wiped clean. In addition, inaccurate heading and speed inputs can cause errors in the receiver's position-fixing computations during satellite passes. An error of 1 knot in the estimated speed over the bottom, for example, will produce an error of as much as one-fifth of a nautical mile in the computed position. On an ocean passage, of course, such an error matters little.

What's Available (And Buying Tips)

Plan on spending $1,000 to $5,000, for which you will get some incredibly small equipment that will do an outstanding job. (The fluxgate compass, if one is available for your receiver, is not included in the basic price.) All the receivers on the market are of good quality, since uniformly high sensitivity is necessary to receive the satellites' one-watt transmissions. All the receivers execute their fundamental task—position-fixing during satellite passes—with efficiency. The less expensive receivers merely have fewer optional features. New, top-of-the-line satellite receivers feature waypoints, anchor watch alarms, and all of the amenities found in sophisticated loran sets. Some satellite receivers will even tie into a loran set to give you additional information between satellite passes, although you will probably be disappointed in the loran output once you get spoiled by the accuracy of satellite navigation. Indeed, some mariners see no need for loran at all. They consider a depthsounder, an RDF, and a radar unit to be a sufficient electronic navigation package for coastal waters, with a SatNav receiver being the next most desirable item.

Choose a system that's easy to understand and operate—and most are. The controls have been made "user friendly" with display prompting—most readouts will ask you what you want next. Unless you have automatic speed and heading input, the readout will also ask you for that information. The amount of information a more sophisticated receiver will display is dependent on its programming. If you elect to program your unit's microprocessor memories to the hilt, it will give you a veritable wealth of output. If you let the receiver function on its own, it will simply tell you your position within a few hundred feet every hour or so, marking the event with a friendly beep.

In most cases, no tuning is necessary on a new receiver, and installation is easy unless you encounter a speed log-receiver interfacing problem. At the manufacturers' request, SatNav receivers are usually repaired by the manufacturers rather than the dealers, at least when the malfunction involves the integrated circuitry. Even if you buy your unit from a specialty dealer, there is a good chance it will be returned to the manufacturer for repair (although the dealer *may* give you a loaner unit to use until yours comes back). For all these reasons, you might consider buying your SatNav receiver from a discounter and saving some money, *if* you have a good idea of what you want.

Installation

Installation of the equipment is more than straightforward—in most cases it's a snap. Noise levels and grounding are not an important matter in most satellite installations, so grab 12 volts wherever you may find it on your main circuit breaker panel. Just make sure that it's a good solid 12-volt source, free from alternator spikes or battery charger hums.

If you have good grounding throughout your boat, run a small ground wire from the chassis of your satellite receiver to any convenient ground point. This cuts down on the amount of radiated microprocessor noise that most receivers generate—noise that can sometimes cause problems in the use of a weather facsimile receiver, ham rig, or marine SSB transceiver. Satellite receiver manufacturers have been slow to admit that their microprocessor circuits are creating electrical noise aboard boats—but most are finding better ways to shield their receivers' innards.

The antenna system is fed with coaxial cable. As with VHF radio and loran systems, more and more manufacturers are enclosing crimped, no-solder coax connectors for do-it-yourself installations. These work well, though they do not last as long as soldered connections. Crimped connections are available separately if they do not come with your unit. Most antenna systems feature a pre-amplifier, similar in style to the loran pre-amp. This pre-amplifier is at the base of the antenna system and picks up its voltage from the same coaxial cable feedline that carries incoming signals from the active antenna element. The antenna is usually sealed in white fiberglass. Mount it as near vertical as possible for maximum reception during low satellite passes. You can, if you wish, hide the antenna belowdecks in a fiberglass hull and still get adequate satellite reception. Mounting the antenna high atop a mast is totally unnecessary, and is really not recommended.

Interfacing the receiver with a knotmeter will be the only possibly difficult aspect of the installation. Here a dealer can be a great help, saving you from possible frustration. If you want the knotmeter interface, and you want to go it alone, you can contact the manufacturer or seller of the receiver and obtain the appropriate interfacing kit. You'd be better off not to go this route, however, if you're not good with a soldering iron.

Global Positioning System

The global positioning system (GPS) may ultimately phase out Transit satellite systems, loran, and even radio beacons. With GPS one can imagine position accuracies, to include altitude, within a few feet. The system was developed by the U.S. Department of Defense to meet commercial, military, and (last on the list) recreational user needs. Although the system is operational, it's far from complete, and the January 1986 Challenger disaster has delayed completion by at least five more years. Coupled with a second-in-a-row failure of a Titan rocket, it seems apparent that the U.S. satellite launching program is suffering.

When completed, the GPS system will include 18 satellites in six orbits at an altitude of 10,000 miles. The orbit period is exactly 12 hours. Your satellite GPS receiver will "see" at least four GPS NAV-STAR satellites simultaneously anywhere on earth at any time of the day or night. A ground control network receives the satellite signals, computes updated orbit and clock parameters, and loads appropriate information into the satellites' memory systems. The satellites continuously transmit navigational signals to the users below.

The seven operational Block I prototype satellites continue to work well and give us approximately 12 hours of navigational information a day in two dimensions, and about 3 hours per day in three dimensions (allowing altitude computations). The signals are sent on a spread spectrum frequency transmission near 1,500 MHz, much higher than the 140 MHz Transit satellite system. This microwave transmitting scheme is of importance to the military because it resists jamming and gives precise information independent of local weather conditions.

The GPS system offers civilian as well as military position-finding data. The military gets extremely precise information, where civilians get somewhat degraded but nevertheless precise fixes, the stuff of navigators' dreams. Receiver manufacturers are providing equipment that gives outstanding accuracy, within a matter of feet. Manufacturers are also combining loran and global positioning systems capabilities, as well as Transit and global positioning systems, into one receiver. This will allow users to upgrade their equipment as the global positioning system is upgraded.

If you're thinking about purchasing a satellite navigation receiver, carefully weigh your opportunities. If you have no navigational receivers at all, you may wish to invest in a single receiver with loran, Transit, and GPS capabilities. If you already own a loran and Transit satellite receiver, on the other hand, you may be more interested in a single-purpose GPS receiver. GPS receivers are still very expensive—in excess of $10,000—but prices should fall in the future.

There is only conjecture as to when all 18 satellites will be in place and operational. Expect it before the year 2000, but don't look for any new developments until 1989.

12

Security Systems

by Gordon West

Boat break-ins and thefts of marine equipment are on the rise, and statistics from law enforcement agencies make it clear that items of marine electronic and fishing gear are popular targets. Because marine electronic equipment has become ever smaller and more portable, it can be removed in a matter of seconds. Trailered boats are particularly susceptible because a burglar has only to unsnap the canvas boat cover, and he is in, hidden from view and able to take his time while methodically disassembling every piece of electronic gear and searching for anything else of value. Anyone who keeps expensive equipment on his boat would do well to consider a security system. Many such systems, specifically manufactured for the marine market, sell for under $200, and experience has shown that they are effective in protecting a boat and its contents.

A word of caution: Most home security systems have no place aboard a boat. For example, inexpensive ultrasonic home security alarms are great for their purpose, filling a room with ultrasonic waves that will trigger an audible alarm if disturbed. Let anyone enter the room, and the alarm is tripped. On a boat, however, use of an ultrasonic system will lead to constant false alarms. A boat cover moving with the wind, the wind currents themselves, hot air funnels, a tinkling of keys, high-pitched disc brakes near the dock, mechanical emergency vehicle sirens—any such disturbance could trigger a response.

Another type of home alarm sometimes installed aboard small boats uses radar waves rather than ultrasonic acoustical waves to guard an enclosed space. In a home, these work quite nicely, even sounding through walls—and there's the rub. The radar waves just as readily sound through a fiberglass hull and can detect the relative motion of pilings as the boat moves up and down, or boats passing by, or other movements remote to the boat itself. After a few of the resultant false alarms, neither the local harbor patrol nor your neighbors will be overly pleased with your new system.

How a Marine Alarm System Works

Probably the best sort of alarm for small boat use—and certainly one of the least expensive—uses small wires to protect everything on board. Two generic varieties of these wire alarms are available—open- and closed-loop systems. Both work well.

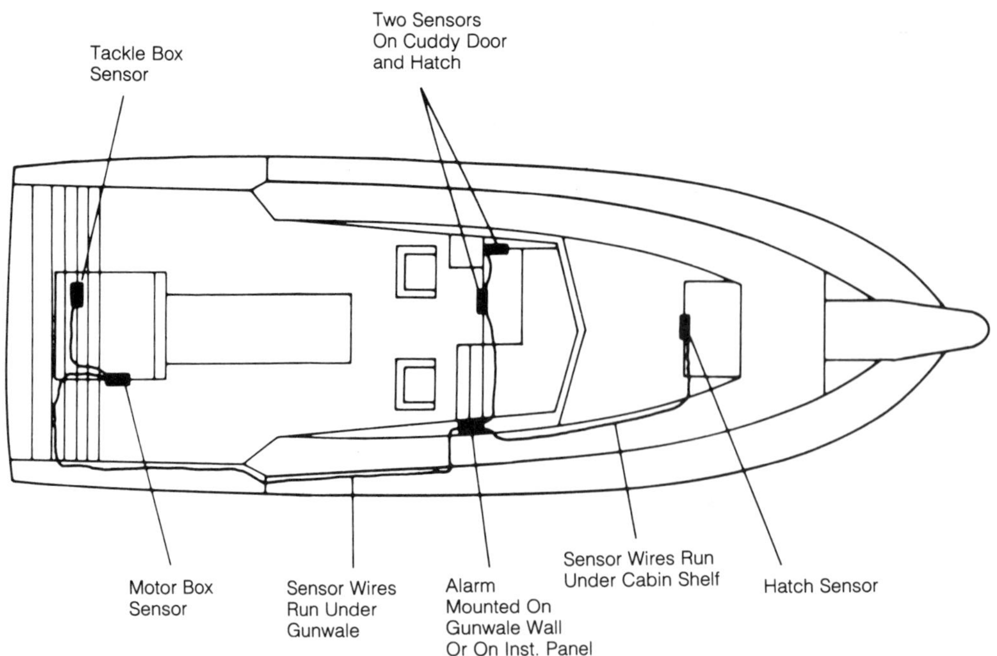

Open-loop system—two wires to each sensor

The open-loop system is normally wired so that the alarm sensors are kept open until something is disturbed. Most manufacturers use hatch-mounted magnets that energize small, wired microrelays. The magnet keeps the relay open until the hatch is opened, at which point the weighted relay snaps shut and the open-loop alarm system is triggered. In one variation, a special sensor snap is used along with the usual snaps on the canvas cover of an open boat. When the boat cover is buttoned up, an open circuit is created; anyone pulling apart the snap while trying to unbutton the cover will automatically set off the alarm system. The sensor snap should be placed in a part of the cover through which a burglar would be likely to enter. Individual instruments can be protected by snap switches if fitted with canvas covers. The switch is mounted on the bottom half of the snap.

Other types of open-loop intrusion detectors include concealed sensors that close a connection should someone walk on a pressure-sensitive mat, and sensors that are triggered when gear is moved out of the magnetic field that holds the sensor open.

The closed-loop security system strings its sensors in series, like Christmas tree bulbs, and will protect everything within this "chain." The closed loop is easy to set up: Pass a single wire through every piece of electronic gear you wish to protect, and terminate the wire at the security alarm master unit. If anyone cuts the wire, the alarm immediately goes off. Alarm sensors that normally remain closed may be wired in as desired. As soon as a magnetized object is moved away from one of the sensors, it will trigger the alarm by opening up a contact.

The merits of open-loop versus closed-loop systems are much debated among security manufacturers; in practical use, both systems work extremely well. In fact, since both are universally accepted, most alarm systems incorporate contacts on the back of the control box to accept either configuration.

With either type of system, the control box must be mounted where no burglar can easily get at and disarm it. Most control boxes allow placement of a remote key switch so that you can turn the alarm off before stepping aboard your boat. In some systems, a remote keyboard with numbered keys that must be pushed in the correct sequence accomplishes the same purpose. The switch or keyboard should be completely weatherproof.

The alarm control center should be wired directly to the vessel's 12-volt battery system, allowing the alarm to function even after an intruder cuts

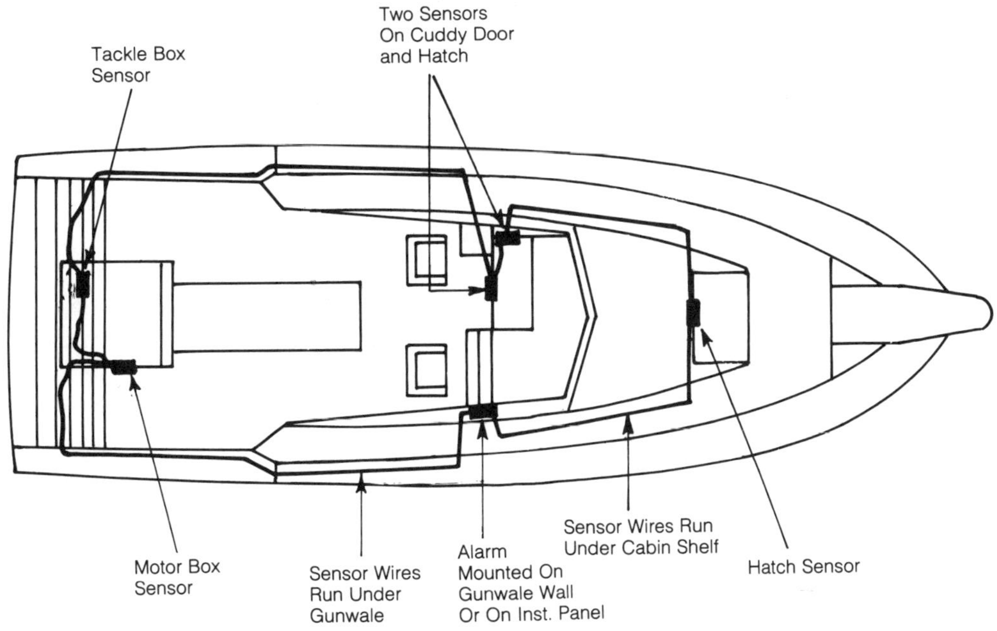

Closed-loop system—sensors wired in series

the 12-volt wires at the fuse panel. Some alarms feature a built-in rechargeable battery that will power the alarm even when the ship's battery supply gets cut off from the alarm box, and a few will sound off automatically if the 12 volts is disconnected. The control system usually features a delayed circuit, allowing you to place your turn-off switch somewhere inside the cabin. This gives you, normally, about 30 seconds to enter the boat and turn off the alarm. If you exercise this option, make sure the control center is well hidden from a burglar to keep him searching a good deal longer than the delay time.

The control center accounts for the big price differences among alarm systems: the more features it offers, the higher the price. Any good marine system should have the following accessories and capabilities:

- Automatic alarm if voltage is removed
- Normally open and normally closed compatibility
- Terminals to enable wiring of a remote outside key control system
- An optional plug in the AC power supply with float charger for a stand-by battery

- An adjustable exit snap bar allowing delayed entry time of 10 to 60 seconds
- Adjustable siren alarm time of 2 to 12 minutes
- Automatic alarm shutdown after 10 minutes of sounding
- Optional outputs to turn on lights, strobes, and auxiliary sirens
- Optional output to a radio transmitter

As the list illustrates, while the basic operation of an alarm system is quite simple, options are plentiful.

Most mariners opt for an audible siren, which is enough to scare almost anyone away from your boat. The siren should shut down automatically after a few minutes of sounding; in some states, this is a law. No one likes listening to an audible alarm for more than a few minutes at a time. Some alarms yield a constant 12-volt output when triggered, permitting a broad choice of sirens (including a voice-synthesized alarm that yells "Burglar, burglar"). Strobe lights are also effective, and assist the harbor patrol in tracking down the boat. Alternatively, some alarms will flash your spreader lights, masthead light, or anything else that takes 12 volts.

Company	Product Name
Marine Alarm Corp. 16 Shore Road Hopatcong, NJ 07843	Boat-Sentry
Marine Security Systems 2999 Pt. East Drive N. Miami Beach, FL 33160	BCI
Marine Technologies 1001 Brentwood Lane Mt. Prospect, IL 60056	Secur-It-Izer
Maritech 1537 "A" Fourth St., Suite 33 San Raphael, CA 94901	Boat Alert
Mooring Products of New England PO Box 200 Waterville, ME 04901	Partner
Aqualarm 1151 "D" Bay Blvd. Chula Vista, CA 90211	Aqualarm
Sentry Devices 33 Rustic Gate Lane Dix Hills, NY 11746	Sentry

Some alarms have dual 12-volt outputs, one to trigger another alarm such as a speech-synthesized horn that yells "Fire, fire." With this arrangement you can wire in fire sensors; the control box will know whether it is detecting a fire or a burglary, and sound the appropriate alarm.

Installation of an alarm system is rendered simple by the manufacturers' suggestions. Again, the control box is usually hidden belowdecks, out of the weather and out of sight. Feed 12 volts directly from the ship's battery to the control box, and keep the 12-volt wiring out of the way, so no one can detect and cut it. (Even when the system includes an automatic alarm when the 12-volt source is removed, visible wiring would help an intruder locate the control box.) Once activated, the system draws only about one-fiftieth of an amp; you could leave the system activated for months on end without pulling down a regular 12-volt battery.

Run normal, two-conductor number 14 or 16 wire to the sensors or through the various pieces of electronic gear to be protected, keeping the wire as neat as possible. You needn't worry too much about whether or not the wires are exposed. Most burglars won't cut alarm wires, because many alarms need to see a normally closed circuit, and cutting a wire could thus trip the alarm. Furthermore, the metal wire cutters used to slice through each pair of cables might accidentally trigger the alarm. Indeed, burglars usually shy completely away from a boat with alarm wires visible.

The idea when arming a boat with alarm wires is to keep every sensor point as simple as possible. Connecting a sensor improperly can result in an inoperative alarm. If your vessel uses a normally open circuit, you should be able to inspect every sensor easily. In a closed-loop system, you get an immediate indication that everything is armed from the alarm status light. The light should come on when you set the switch. A great advantage of the closed-loop system is the ease of doublechecking that everything is indeed activated.

When the system is installed, have a friend try it out. See how long it takes him to set it off accidentally, and how long it takes to find the source of the noise and disable it.

If you use a siren, make sure that it can't easily be tossed overboard or otherwise deactivated. A plastic siren that makes a lot of noise inside a cabin can immediately be silenced by ripping it from the wall and breaking off the wires. Locate an alarm horn or siren in a hard-to-reach place.

Police reports indicate that most thieves will immediately high-tail it if they hear an alarm that they can't easily disable. The reports also indicate, however, that cheap alarm systems are immediately detected and destroyed by a thief, who then goes about his business.

13

Radar

by Gordon West

At night or in the fog, marine radar will show you where you are, where you have been, and where you are going. It will let you spot surrounding dangers—buoys to starboard, the jetty to port, or other boats around you in the thick of the night or the height of the storm. More radars have been installed by sail and powerboat owners during the past few years than ever before, convincing evidence that the advantages of radar for piloting and safety are now being recognized by an increasing number of mariners.

For years, radar (an acronym that stands for RAdio Detection And Rangefinding) was considered an expensive proposition—cumbersome to install, heavy on the electrical system, and difficult to use. In the last five years, however, thanks to foreign manufacturers, radars have made a quantum jump in technology (and popularity) with price levels remaining almost the same. Quite simply, you now get almost twice as much radar for the buck as you did five years ago, and your boat's electrical system will love it—the new sets draw about half the power of the older ones.

How It Works
And What's Available

Marine radars work on essentially the same principle as echo sounding. Radar waves are timed as they leave the antenna unit, bounce off targets, and then re-enter the antenna system, where they are fed into an amplifier and then the display unit. Unlike the depthsounder, which points only downward, radar signals are beamed out steadily, sweeping a continuous circle and thus giving direction as well as range. Since radar radio waves are transmitted on frequencies much higher than our marine VHF band, the signals travel line of sight (with a 20 percent refraction over the horizon). At such a high frequency, radar waves are very reflective, as they need to be.

Inside the radar is a timing device that measures time delays in microseconds (millionths of a second). At close to the speed of light, a radar pulse will travel 328 yards in one microsecond. If one microsecond passes from the time a pulse is transmitted to the time its echo is received, that burst of energy has traveled 328 yards—164 yards out to the target and 164 yards back. The radar displays the target distance as 164 yards from your boat's present position. Antenna position information is also relayed to the display unit, so that not only the range but the precise bearing of the target is displayed on the scope. Finally, the signal strength of the target on the scope is an indication of its size and density.

The basic components are the antenna unit, the transmitter and receiver circuitry built into the antenna unit, and the display unit, which contains the timing and computer circuits, the cathode ray tube, and the power supply. The transmitter, receiver, and antenna system are housed in a 30- to 60-pound package, and the display unit is a lightweight, compact unit that mounts below. A one-inch-diameter cable connects the antenna unit with the display unit, and the display unit gets its power from a relatively small pair of 12-volt wires.

Inside the antenna unit, the transmitter section develops powerful energy pulses on a frequency near 9,400 MHz. The length of each pulse of energy might be as short as 0.05 microsecond for short-range calculations or as long as 1.0 microsecond for long-range calculations. Shorter pulses at closer range give better target definition and resolution. On the short-range scale, the pulse repetition rate is also increased, to as high as 2,000 to 2,500 pulses per second. This allows the receiver to listen for echoes between each transmitted pulse. At longer ranges the echoes will take longer to come back, so the radar will automatically slow down to approximately 800 to 1,200 pulses per second with pulse lengths many times longer. Again, the radar must immediately switch to receive after each high-powered pulse is transmitted in order to allow the receiver section to pick up the echo. A modern, microprocessor-based radar automatically selects the optimum pulse repetition rate and pulse length depending on which range you have selected.

Most modern radars offer a minimum of five selectable ranges, the shortest of which may be 1/4 mile, and the longest, 16 to 20 miles. There *are* radars with more power and longer range, but long range is the *least* important feature for a small-boat radar system. Since radar waves don't bend over the horizon, it makes no sense to buy a 40-mile radar when you can only mount it high enough to put the horizon 8 miles distant. If you are planning to operate along coasts with high mountains close by the shore, you might choose one of the longer-range radars. Remember, though, that for a radar to detect a target at a maximum range of 20 miles, your radar antenna must be 10 feet or more above the water and the landmass must be more than 170 feet above sea level.

Small-boat radars normally transmit with about 3,000 watts of peak power in every pulse, while the receiver section is capable of detecting signal echoes as low as one-tenth of a microvolt in intensity. The antenna completes 30 to 40 360-degree rotations per minute.

The antenna unit may be encapsulated under a fiberglass radome, or it may be out in the open. Choose the former if you are going to be mounting it aboard a sailboat halfway up a mast, since sails and rigging would foul an unenclosed antenna. In either case the assembly consists of a horizontal reflector that beams out and retrieves the radar pulses and echoes. Beam widths may be from 1 1/2 to 5 degrees depending upon the width (or span) of the antenna. Open antenna reflector units are commonly 3 feet long, but a longer unit (4 feet, for example) will concentrate radar energy in a tighter horizontal beam, giving you greater target definition. The small antenna assemblies enclosed in radomes, while they are very compact, might show two fishing boats close together at a range of 5 miles as a single echo on the scope, while the slightly longer, open antenna would show the vessels as two separate targets. Longer antennas lead to better target definition.

The antenna assemblies are designed to offer reliable operation even when your vessel is rolling, the vertical beam width of most radar signals being greater than 25 degrees. In heavy seas, however, a distant target might "disappear" in the trough of a wave at the same time your radar sweeps that sector. Intermittent targets in heavy seas are probably small boats a few miles away.

You may wish to mount the antenna unit as high as practicable to increase the maximum realizable range, but remember that while it might be nice to have the antenna halfway up a sailboat's mizzenmast, it's not really necessary for good coastal navigation and short-range ship detection. Antennas just 10 feet off the water will give you a good view of breakwaters, ships, and any other targets within 4 or 5 miles, and it's what is happening in your immediate neighborhood that concerns you most.

When an antenna is mounted down low, there is always some concern over the potential harm from radar waves. Since no type of radiation can be beneficial to the body, it would certainly be best not to have the radar setup just a short distance away and sweeping by at eye level. Many installations on powerboats do put the antenna assembly directly in front of the operator on the flying bridge, and there have been no reported adverse effects. Probably the

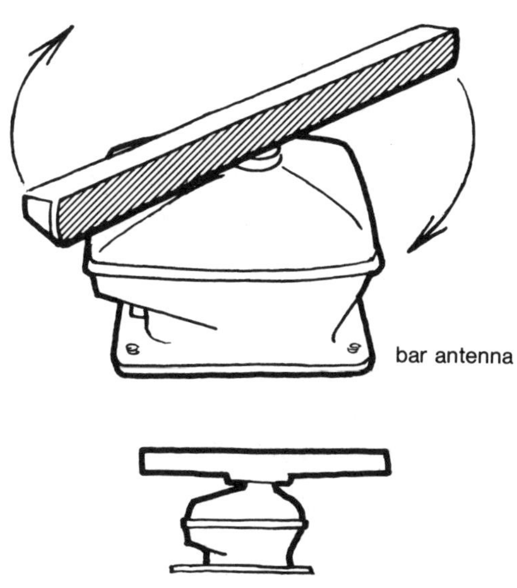

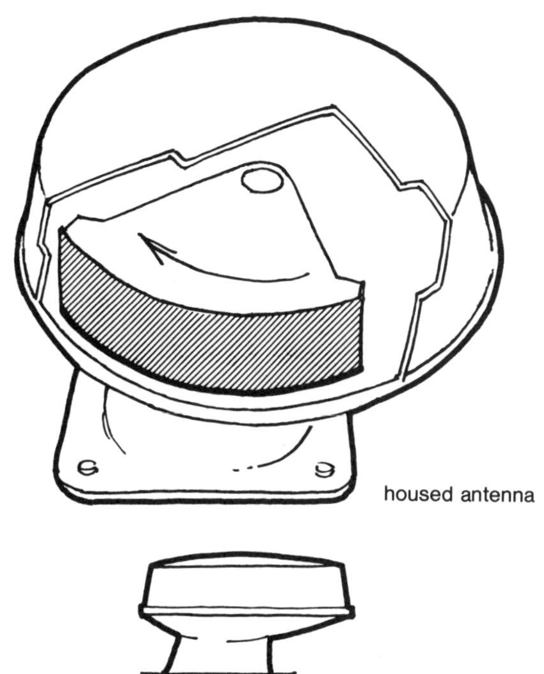

Radar antennas may be of the open type (**left**) *with a span of 36 inches to as much as 9 feet, or enclosed within a radome* (**right**). *In the latter, the span is seldom more than 30 inches.*

The antenna systems used on most small boats have the transmitter and receiver section with its antenna rotating motor in a watertight housing directly below the antenna.

intermittent sweeping of radar waves is no more harmful than standing beneath a large commercial broadcast station antenna tower setup. The sensible approach is not to court trouble: There's enough radiation in the slotted antenna reflector to heat up a bologna sandwich. Mount the antenna assembly well above all occupants on the vessel and you can put your mind at rest.

The display unit is the component that has seen some dramatic improvements. At 20 pounds or so, it's quite lightweight, and the internal power supply in the indicator unit is fully transistorized and uses new circuitry in place of heavy transformers. The unit also houses the microprocessor electronics that allow you to select the many different modes of operation for the radar. Most small-boat radars will "see" as close in as 30 yards from the vessel when on their 1/4- or 1/2-mile range scale, and most have five or more selectable scales for short-range, medium-range, and long-range operation. When

shopping for a radar, note how many short-range scales each radar offers for detecting targets less than 5 miles away. Again, what is happening close by your vessel in the night or the fog is your primary concern.

A radar set offers other controls besides the range selector, of course. A rotatable cursor allows you to display a straight line that emits from your location in the center of the screen. You can rotate the cursor to put it on line with a target, and the target's relative bearing may then be read where the line intersects the 360-degree azimuth at the scope's perimeter. With this feature, you can plot the bearings of other vessels in your vicinity, and watch their movements. If another vessel draws progressively nearer along the cursor line, you are on a collision course. Some radars also feature audible anticollision alarms that will be triggered when an approaching target closes within a preset range along an imaginary cursor line. Alternatively, an alarm can be set to go off whenever any target, regardless of its relative bearing, is detected within a predetermined range. If you get too close to shore, for example, the alarm will be triggered, calling your attention to the scope.

The electronic variable-range marker is another relatively new refinement in radar display consoles. Instead of having to estimate the distance to a target by interpolating between the fixed range rings that match the range scale you have selected, you can electronically expand or contract the circular range ring to coincide with the distance to a target. The range (to as close as one-tenth of a mile) is displayed digitally.

The traditional radar scope utilizes a cathode ray tube with targets and range ring information displayed as bright orange "returns" on the plan position indicator (PPI) scope. Your vessel is located in the center of the picture and emits a faint scanning line that sweeps around the scope every couple of seconds, in synchrony with the antenna rotation. When this scanning line detects incoming signals, it literally paints a picture on the scope showing the bearing and range of the target and its relative intensity on that sweep. The targets are re-illuminated with each rotation of the antenna. Fast-moving targets may appear to leave a phosphorous "wake" as you watch their progress across the screen.

This display system is still used on most radars today. An initial inconvenience is that when you make a tight turn, the shoreline "smears" as old targets decay and new targets appear in different locations. After a while, however, you will get used to this, and you will experience no problem interpreting the radar in heavy seas, tight turns, or simply while plotting fast-moving vessels across the scope.

The only real drawback is that you can only turn up the intensity of the display so far. Most presentations are orange and are easy to see in a reasonably darkened pilothouse. When a cabin admits a great deal of light, however, a hood is required for daytime viewing. It is possible to purchase a hood with a corrective lens built into it to compensate for near- or farsightedness. Magnifiers may be ordered with a radar setup to increase the relative size of the PPI scope presentation. The conventional PPI scope has been around for years and will continue to serve the industry for many years to come.

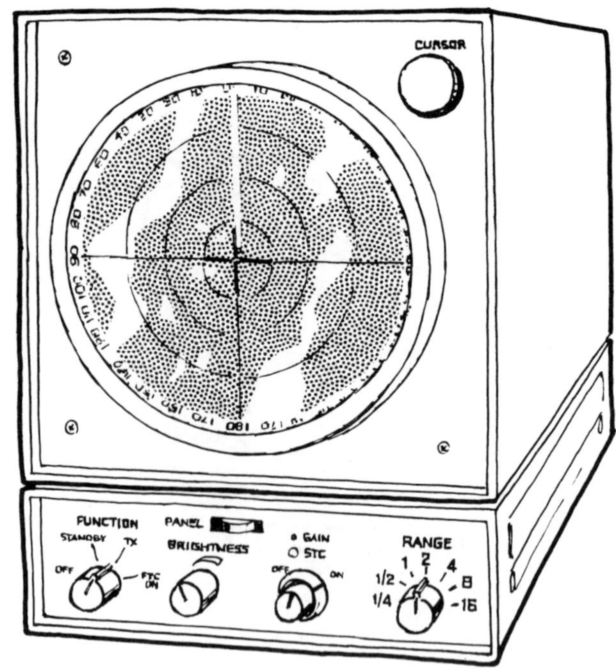

The number of controls on a radar display unit varies with the manufacturer and the radar's maximum range, but every model must have at least four, with one or more serving two purposes. At the upper right corner of this display console is the cursor rotating control for moving the cursor line through the full 360 degrees to obtain the relative or magnetic bearing of a target or headland; at the bottom are the controls to energize the radar and adjust its gain and sensitivity, its range scale, and the brightness of its display. The price range of a new radar (with antenna) is $1,500 to $5,000, and the size of a display unit like the one shown here might be 10 inches, by 10 inches, by 15 inches.

Recently a new type of display has gained in popularity. Called "raster scan," it looks a little like a video game. The raster scan display gives you instant, digitally processed echoes that may be displayed in sharp contrast to areas on the screen where no echoes are present. You can even choose which color you want the echoes to be. Some models have a white screen with black echoes or a black screen with white echoes. Others offer an orange background with green targets.

Each echo consists of tiny squares, the number of the squares corresponding to the intensity of the echo. Weak echoes may only illuminate two or three squares, while large echoes may illuminate a block of 20. As many as 250 to 500 lines per side of the screen are used to define the squares—a long way from the resolution and sharpness of a television picture, but effective nevertheless. On a conventional phosphorous PPI scope, a gently sweeping shoreline would appear as a smooth curve. On the raster scan, it would resemble a million little bricks comprising a jagged, stepped outline. After a while, however, one can get used to seeing blocks rather than smooth curves.

The echoes on a raster scan display may stay illuminated for several seconds, even though the microprocessor hasn't seen their target return for a couple of sweeps. Then, suddenly, the echo disappears. This, too, you must adapt to if you have previously operated phosphorous-type scopes, in which targets fade gradually.

Some manufacturers have followed in the steps of video depthsounder makers and have assigned different colors to different echo intensities. This may help the mariner identify extremely strong targets that grow in intensity but barely swell in size on the raster scan display.

Probably the biggest benefit of the television-type display over conventional phosphorous PPI scopes is the fact that it is inherently brighter and easier to see in a bright pilothouse. You may never need to use the hood in order to create shade around the scope. Another benefit is that you can add more than one display unit to your system. Once you set the controls down below, you might have a number of additional readouts—each remote readout simply hooking up to 12 volts and a common coaxial cable feedline, just as in cable television. This can't be done with the PPI phosphorous scopes.

It is to be hoped that raster scan, as it continues to evolve, will give us more lines per inch for better resolution. Look for this to happen, and if you're shopping for a raster scan, use the degree of target definition in the display as one of your criteria.

Regardless of the display you choose, you needn't worry about power consumption. At 12 volts, some 20-mile radars pull less than 4 amps. Since the engine is usually running when the radar is in operation, you should never have a problem with the radar draining your batteries.

Operating Procedures

Before turning on your radar, doublecheck your marine radiotelephone license and see whether or not it has been endorsed for 9,400 MHz (9.4 GHz) radar operation. If it hasn't, you must fill out a new Form 506 and specifically request radar frequencies from the Federal Communications Commission. As soon as you drop this form into the mail, you can start operating the unit. When your license comes back, you will have the same radio call sign, but with additional frequencies allocated for radar use. You are not allowed to operate radar without the FCC frequency endorsement on your marine radiotelephone license.

Begin using your radar set in the daylight and note how the objects around you appear on the screen. You will notice that radar doesn't see over large ships or around islands, nor will it pick up low-lying rocks at distances greater than 3 to 5 miles. Radar works on a line-of-sight basis.

The type of target will influence its ability to reflect radar signals. Sailboats with sails set are poor radar reflectors. Radar waves simply penetrate sails, and an aluminum mast reflects radar signals in scattered directions. The effect is akin to bouncing a tennis ball off a round pole.

Some targets will make huge presentations on your radar scope. Supertankers turned broadside are an example, as are buoys with radar reflectors. Anything comprised of flat metallic surfaces that intersect at 90 degrees will reflect radar signals with great intensity. This is the benefit of mounting a radar reflector on a vessel.

Since fiberglass is a poor conductor except when wet, fiberglass boats are not good reflectors of radar waves. You will notice, too, that different shorelines give readouts of different intensity. Low, sandy beaches barely show up until you are almost on top of them, while sharp cliffs reflect signals with tremendous intensity. Low, rolling hills give only fair echoes, and sandbars will give almost no return echo at all.

Heavy seas may reflect radar signals, causing confused readings at times. There are controls to help eliminate sea clutter return, however. You can also spot approaching rainstorms with radar, or manipulate the appropriate control to help eliminate rain return.

How sensitive is a radar? A good radar set should pick up a low-flying pelican up to one mile away, or a seagull resting on flotsam one-quarter mile away. The calmer the seas, the higher you can turn up the intensity control for maximum sensitivity and target detection on a phosphorous scope radar.

Once you have gained confidence in detecting targets, the next step is to learn to use radar to avoid groundings and collisions. If you carefully align the cursor over a target, you can watch its progress and ensure that it is not on a collision course with you.

You can safely navigate within a harbor using the radar on its shortest range setting. Enter the harbor a few times in the daylight with the radar set on, and you should be able to rely almost completely on radar for a nighttime approach. Consult your chart for the buoys that appear as targets on your scope as you approach the harbor. Watch the shorelines or breakwaters as you pass through the entrance, and then look for small craft that might have been "hidden" behind a breakwater. As your skill increases, you'll be able to spot and negotiate the appropriate channel in a crowded anchorage, coming within 30 yards of your dock on the basis of radar alone.

At sea, you can use a radar for position finding by measurement of distance and bearing. Spot large mountains on your charts, and locate these landmarks on your display, with the radar set to one of the longer ranges. Distant peaks will appear as small traces toward the scope perimeter. Use the cursor to develop lines of position.

By extension, radar can also be used to determine speed over bottom or vessel drift. It will even tell you when other vessels in the area are using radar, since curved interference traces will appear on your radar with every antenna revolution. Some mariners may curse at these interference traces, but they are really quite helpful. They can, however, be eliminated with special circuitry.

The United States Coast Guard has made it a bit easier to identify certain targets by equipping them with racon devices. Assume you are entering a harbor and need to ascertain which target is the main channel buoy; you'll be in luck if that buoy carries a racon device. It will paint a coded signal on your radar set, removing all guesswork. In the manner of the air traffic controller "squawk" signals that identify aircraft, racon buoys and shoreside racon installations mark the exact spot of a reflected signal. You can expect to see more of these devices in the near future, as radars become more popular.

Additional Buying And Installation Tips

You must really see a radar set to judge it, and for that reason, particularly if you do not plan to install the unit yourself, you would be wise to purchase it from a specialty marine electronics dealer, one that is an authorized sales and service agent for that particular unit. You'll want someone local in case the radar should require service, so that a technician can be brought aboard for the work. Most of the electronics are in the antenna unit, which is not easy to dismount for repair off the boat.

Radar sets require high voltages, and high-voltage circuitry is inherently more prone to malfunction. This is another reason why most radar sales take place at the dealer level rather than through mail-order sources. It takes a skilled technician at a hefty hourly rate to work on your radar, but if you develop a service contract with the marine electronics dealer that sells you the equipment, you can save yourself some money down the road. Most radars require service at least every two years due to their high operating voltages as well as the sensitivity of the magnetron transmitter output tube. If you use your unit a lot, it will require annual peaking and tweaking. For best results, this can only be done right on your boat by an experienced technician. Buying from a specialty dealer— and insisting on a firm service contract—is the surest way to a long and happy relationship between you and your radar.

If you *do* decide to install your own radar, follow carefully the instructions in the manual that comes with the unit. Unless the manufacturer includes an FCC certificate in the packing case, it will be necessary to have a technician "sign off" your installation and make an entry in your radiotelephone and radar logbook as required by the FCC. If you buy the unit through a mail-order house, be prepared for this added, not-insignificant expense. The local marine electronics specialty dealer from whom you *didn't* buy the unit may not be eager to send down a technician to sign you off for just a few bucks. Indeed, he may charge you several hours of service time for this sign-off procedure.

A few marine electronics mail-order companies will certify a radar before it is shipped to you. If this can be done, it will save you some hard-earned money once everything is aboard and operating. Radars are so sophisticated and self-adjusting that a do-it-yourself installation is feasible *if* you are good with tools and can devise an appropriate antenna-system mount. Some manufacturers offer pedestals for mast-mounted antennas, and there are countless other possible approaches. Remember, though, if you're going to mount the antenna high above the deck, that there may be tremendous stresses on the installation when your vessel is pitching in heavy seas. Since the antenna unit contains almost two-thirds of the radar's electronics, make sure it's bolted down firmly. Replacing it would be expensive.

The antenna assemblies are matched for the radar set you are buying. When you buy the radar, you buy it as a complete setup, so you needn't worry about matching the antenna assembly to the radar. If you're given a choice of antenna lengths, try to accommodate the slightly longer antenna for better target definition. The cable connections between antenna and display unit are straightforward; you can hook it up yourself simply by following the directions and screwing in the wires by the numbers.

Prices for radar setups have been holding steady for the past two or three years at $1,500 to $3,000. Don't look for any dramatic price drops in the near future. Thanks to foreign competition, however, more and more features are being made available for the money. This trend is likely to continue.

Maintenance

Keep the display unit out of the weather to protect delicate circuitry within the console. Don't leave it in a damp flying bridge for any length of time. If you're going to be storing the boat for some time, take the display unit home to store in a dry place. Never seal it in a plastic bag—that only collects moisture.

Now and then, give your antenna radome a good bath with warm, soapy water. Radar waves have a hard time sounding through accumulated soot. Never paint the antenna dome—lead-based paints, especially, will literally soak up radar signals. The painting of fiberglass radar antenna domes is a common practice, but a very poor one.

Never tune the intensity up too bright. It should be just bright enough for viewing. The phosphorous cathode ray tube displays, especially, are easily damaged by a too-high brilliance setting.

14

Autopilots

by Freeman Pittman

Autopilots are inventor Elmer Sperry's gift to plea-sureboaters—or they are a pain in the neck that don't work when you really need them. Skippers' reactions to them are based on experience, and when the system is a good one, it's not easily for-gotten. The sensation of feeling a boat moving surely and economically with no one at the helm makes you glad to be born into this technological age. When the experience is a bad one, you remem-ber that, too. A good deal of the resulting unhappi-ness often stems from not having known enough about what you were getting into before purchas-ing and installing the hardware. The autopilot is one of the most complex systems on the boat and takes up a good deal of the available space and power. Discovering the strengths and weak-nesses—as well as the idiosyncrasies—of each model that interests you will go a long way toward ensuring the reliability and good performance of the pilot you finally choose.

Autopilots have grown immeasurably in sophis-tication in recent years, and yet the most basic design of more than 15 years ago—the Tillermas-ter—is still being marketed, and offers dependable, efficient operation for day trips or crossing an ocean singlehanded. The least expensive electric pi-lot on the market costs $350, much less than a loran and about the same price as a handheld VHF, a cost many small boat owners can justify.

On the other hand, a thorough, permanently installed belowdecks system, costing several thou-sand dollars, can keep you on a compass or wind-vane-directed course with an accuracy of plus or minus 1 degree in good weather, compensating for yaw, leeway, an imbalanced sailboat helm, and even (when a loran or satnav is tied into the system) current set and drift. And it will do all this while using little more of a sailboat's precious electrical power than does a cabin light. Powerboat autopi-lots, by the same token, today deliver more power from lighter, more compact systems than ever be-fore. Trim, torque, or poorly synchronized engines can be offset by these pilots, too.

A chart comparing manufacturers' specifications and prices appears in Appendix C.

How It Works

Years ago the crudest autopilots simply zigzagged the boat across a chosen course, using electrical contacts touching the heading sensor's compass card as the limiting switches. Because of their behavior, these basic pilots were dubbed "hunting" types.

The improvements that have made autopilots viable for small craft have been the miniaturization of components and the development of highly efficient, solid state electronics. The first big improvement was the incorporation of a "deadband" in the system. The deadband, whether mechanical or electronic, is the narrow segment of a compass down the middle of which the boat's course is set. As long as the boat's heading stays within the deadband, say 2 degrees on either side of the course, the autopilot takes no corrective action. In a manual steering system, the total play is its deadband. Wiggle the steering wheel a half-inch or so and you may feel no rudder movement. The boat is free to wander within that range.

Obviously, allowing the boat some free rein saves energy. Within the typical 1- to 5-degree deadband width current autopilots allow, the boat, if it has been trimmed and balanced, will come back to course a fair amount of the time. Making the pilot work only outside the deadband lessens the amount of electrical power needed to drive the motor that turns the rudder.

A higher level of sophistication, if not necessarily a more cost-effective one, is reached by "proportional response" autopilots. Proportional response means the rudder is turned at a particular angle and rate for a given course error. Less rudder is needed for a small error, more for a large error. The proportional pilot applies the right amount for each, in contrast to simpler systems which react with the same amount and speed of helm correction for all course deviations and conditions.

Heading Sensors

Regardless of the principle of operation, an autopilot works only on the command of its lookout. For the powerboat, and generally for a vessel under sail, too, the lookout is the "heading sensor." For many years, autopilot systems have incorporated a modified card compass as the heading sensor. An electrical "pick-up" coil placed above or below the compass card is aligned with the field created by the compass's magnets. A low voltage is fed into the coil. When the coil (and boat) rotate out of alignment with the surrounding field, a change in voltage is created. This signal travels to the control unit for amplifying and processing, and is passed on to the drive unit. Robertson and Wagner are two manufacturers who offer the pick-up coil sensing system. They can supply a compass with coil, or the coil alone for modifying your boat's steering compass.

Another effective sensing system uses a photocell to detect a light beam that passes through a compass card divided into clear, partially shaded, and opaque sections. To set a course, you line up the boat and rotate the photocell and light around the compass so the photocell beam passes through the translucent segment of the card. If, through rotation of the card due to course deviation, either the opaque or clear portion swings into the path of the light beam, a signal change is generated that activates the pilot's drive motor. The opaque portion cuts off the signal and makes the pilot turn one way; the clear portion allows a full-strength signal that makes it turn the other way. Cetec Benmar and TillerMaster have used variations on this sensing system successfully for years.

Both the pick-up coil and light-sensing systems are repeaters; that is, they follow the movements of the boat around the compass card. Many manufacturers today are turning to a fully digital heading sensor that replaces the card compass itself. This is the "fluxgate compass." You don't see numbers on a card with the fluxgate; instead, its inner works resemble more the antenna of a tiny radio direction finder. It works much like an RDF, too. A radio signal is strongest when the RDF's bar antenna is at right angles to the signal, and weakest when it is end-on to the signal. For the fluxgate, consider the earth's magnetic field as the radio beam, and use two bars crossed at a right angle as the antenna. Pass a low voltage through coils wound around the bars, and watch the digital readout as the earth's field distorts the coils' voltage output. The amount of disturbance varies as the fluxgate is rotated through 180 degrees. The output phase changes, too, and sensing the phase eliminates reciprocal angle ambiguities. A given voltage output and phase correspond to a particular compass heading. Most fluxgates have a digital LCD readout, allowing them to be used as an alternate to the ship's steering compass. A visual display is not needed for autopilot work, though, since the sensed signal need only be transmitted to the autopilot control unit for amplification.

The fluxgate system is not subject to the oscillations and overshoot of a card compass with an electronic repeater. It also can use less power. Some fluxgate compasses perform better than others: The universal joint from which the wire-wound element dangles differs from brand to brand, and some hang up more easily in a seaway. Damping also affects accuracy, which can be as good as 1 to 2 degrees with the best models. The fluxgate would appear to be more vulnerable to temperature variations and to dip error than a card compass, which can be weighted to deal with dip in higher latitudes. The card compass also remains functional as a non-electronic indicator should the boat's power go off. With a fluxgate, you'd better have a mechanical backup.

A fluxgate compass hooked to a Magnavox SatNav receiver has a unique advantage over any other heading sensor. The SatNav has a program that corrects for both deviation and variation. All you have to do is initiate the program and steer the boat in a slow circle. The SatNav does the rest.

How expensive is the fluxgate, relative to compass repeaters? Robertson-Shipmate has switched its standard heading sensor from a compass with pickup coil to a KVH fluxgate compass. Prices for the two types are close (both around $800), with the fluxgate slightly less expensive since Robertson orders it in quantity. This may not be true for other brands.

Steering a sailboat need not be by compass alone. The sails and helm can be set to hold a course relative to the wind's direction. Some autopilot manufacturers offer a windvane option that can take over from the heading sensor when you're sailing on the wind. This is simply an electronic apparent wind angle sensor like the indicators in many sailboat instrument packages. You set up the boat so the rig and helm are balanced, and dial in the relative wind angle you want to sail. The wind direction sensor takes over from the compass sensor and the boat will stay on a heading relative to the wind, rather than a compass course.

If the windvane is well-integrated with the autopilot logic, it is a feature worth having, especially when you are sailing on the wind while crossing an ocean, and must take time off to rest. But not all windvane-steering autopilots are created equal. An experienced transatlantic sailor says he has not yet found a windvane system that reacts quickly enough to wind shifts. It is also possible for a system to react too quickly, turning the boat without waiting to see if the wind shifts back to its original relative angle. The key here is how often the pilot samples wind readings, and how much it averages them before acting on them. Having a manually adjusted averaging rate might solve the problem momentarily, but since wind strength and direction are always changing, the right rate now might be the wrong one in five minutes. The Autohelm line of pilots now incorporate a different solution to this problem. In these pilots the windvane does not totally pre-empt the compass, but only updates it from time to time.

The Autohelm, the Navico line, and some other pilots employ a windvane on a shaft that can be clamped to the stern rail when in use, and stowed below at other times. Some pilot systems can also be tied into the electronic wind indicator instrument mounted on the masthead. Although you might think this would offer cleaner air and more accurate readings, there is far more motion aloft, and the wind indicator can send the pilot steering data based on wild gyrations. Most of the time, wind sensed from the weather side of the stern gives the best steering results.

Controls

A heading error signal from the compass or windvane sensor goes to the autopilot's control unit. There it is processed and amplified in order to command the actuator motor that drives the rudder. The control unit may be better envisioned as a control group; it performs a number of functions, and not all its components necessarily reside in one housing. The amplifier in one manufacturer's system is in the actuator motor case. Several systems put the buttons and knobs in a compact control panel that's easy to locate in the nav or helm stations, while placing all the microprocessors, switches, and other electronic components in a separate box that can be mounted in the location best protected from interfering electrical fields. However the parts are packaged, the control unit in a permanently mounted system contains an amplifier, logic circuitry, a junction box for all the plugs and connectors going to and from other components, as well as the dials, buttons or knobs with which you operate the system. Portable cockpit-mounted pilots may be completely self-contained, with compass sensor and drive unit included, or they may have the compass in a separate case you can mount belowdecks.

Unfortunately, you cannot rely on the manufacturer's literature to help you understand what all the control unit functions are, since the terminology is not yet standardized. The following explanations should help to resolve confusion.

Yaw (variously called sea state, sensitivity, response, or deadband) sets the deadband width on pilots using that type of control. In other words, it determines the amount of deviation from course (the yaw angle) the pilot will allow before making a rudder correction. The control may be a mechanical one inside the pilot that you set once for your boat's characteristics, or it may be a dial adjustment you can make at any time to suit the sea conditions. A wide deadband lets the boat react to momentary influences such as individual waves in heavier conditions. A narrow deadband keeps the boat on a straight course in smooth water. Given heading sensors with plus-or-minus 1-degree accuracy in smooth conditions and drive motors able to make small corrections with minimal time lag, today's autopilots offer a deadband as narrow as 2 degrees (1 degree to either side of course) and adjustable to as wide as 5 to 8 degrees. The most automated of autopilots do away with deadband control by feeding a nearly constant stream of small power pulses to the drive unit, instead of fewer but larger bursts in response to wider yaw angles. For these pilots a low-pass filter, which won't allow momentary errors from individual wave impacts to be signaled to the drive unit, offers the damping inherent in deadband systems.

The gain control (rudder, ratio, P factor) adjusts the proportion of rudder response to course error. For a given error on a light sailboat with a detached spade rudder, or on a planing powerboat, the gain might be set low, since little rudder movement is needed to make the boat change course. On a heavier, long-keeled boat with an attached rudder, or on a trawler-hull powerboat, more helm might be needed to make the boat respond. In that case, a higher gain setting would make the rudder turn

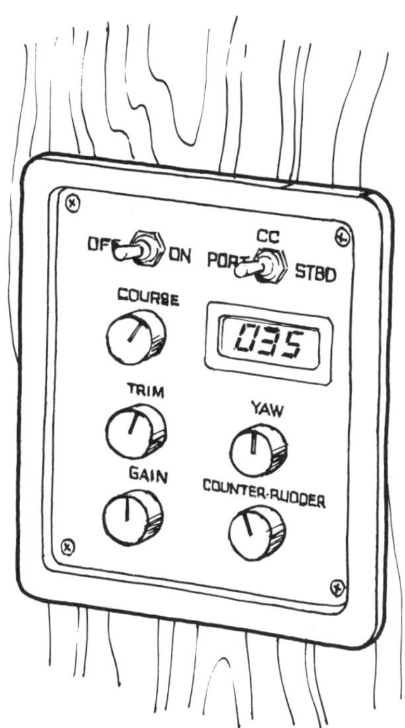

Permanently installed autopilots often have unobtrusive control panels you can mount flush with a bulkhead or dash. Although this unit has about all the adjustments you could want in a good unit, the functions of trim, counter rudder, and gain are now being taken over by microprocessors.

farther for the same degree of heading error. In a proportional-response pilot the gain may also control the speed, or rate, at which the rudder is turned in relation to the rate at which the boat falls off course. From this relationship comes the term proportional rate.

Essential to a proportional-response system is an indicator that signals rudder position to the control unit. On most proportional-rate models the rudder-feedback unit is a potentiometer (the rheostat for adjusting a dining room light is also a potentiometer) attached to the rudder by a small tiller and tie-rod; the potentiometer feeds a higher or lower voltage to the control box depending on whether the rudder is hard over or nearly centered. Recently, some of the newer, solid-state pilot systems attempted to eliminate the potentiometer with its corrosion-prone contacts by including a circuit that simulates rudder angle signals. Such a circuit senses the strength and polarity of signals sent to the drive motor. Comparing these values to those assigned to a neutral helm, the circuit determines rudder angle. Although this system may be more reliable mechanically, it has not proven as accurate as the potentiometer. Now manufacturers like Autohelm and Robertson offer a mechanical feedback unit, and tout the electronic angle simulator as a backup feature.

Counter rudder, a function increasingly being automated, is tied to the amount and/or speed of rudder response. Counter rudder is designed to keep a pilot from overshooting the course while making a heading correction. Some pilots have a manually adjustable counter rudder function to cope with various sea and load conditions. Others are set once for a particular boat's characteristics. The most up-to-date microprocessor models have automated counter rudder that is proportional to the error, just as the rudder response is. In most cases (except when the error is a very large one), the pilot reverses the drive motor long enough to keep the boat from crossing the course line when it is returning from its off-course angle.

Trim (bias, offset) compensates for a persistent heading error, such as from extreme weather helm, only one of two engines working, or a strong, steady wind blowing the bow off to leeward. Some pilots have a trim dial which alters the control unit's sense of proper heading to a position slightly off center. It makes the unit think a slightly turned rudder is actually amidships. You make the adjustment based on visual references outside the boat, or on experience. You can't feel the tiller or wheel for wind or for weather helm, because the pilot has taken over. Some pilots, like the Alpha Marine 4000 series, Autohelm's 6000, Benmar's Compu-Course 2100, and Cetrek's 7000, have an automatic trim circuit that deals with accumulated error by sampling errors over a period of time, and applying the rudder if most of the errors have been to one side of the course. This is a periodic adjustment, not a continuous one like manually selected trim function.

Trim corrections cannot deal precisely with the effects of current- and wind-induced leeway. The boat may keep to its set heading, yet still drift off track. To cope with these influences, an outside position-finding device such as a loran or SatNav must be incorporated into the system. Many permanently mounted autopilots have a loran interface as an option. And because the industry generally has agreed on a standard interface (National Marine Electronics Association 0180 or one of its successors), many lorans, and some SatNavs, can be plugged right into the pilot. The loran or SatNav's continual position updates are compared to the chosen course, and the pilot is temporarily directed to take a heading that puts the boat on track. This is a nifty feature, but beware: Not all so-called compatible navigators and pilots really do interface. Check with the manufacturers of the equipment you're considering to be sure their products are on speaking terms.

The 0180 interface is a simple format which allows a loran's cross-track error computations to be converted into heading-to-steer commands that turn the boat into the wind or current enough to keep it on track. The later, more sophisticated 0183 format, unfortunately, is so broad that few manufacturers can agree on which of its outputs to use for autopilot integration. SatNavs such as the Magnavox 4102 use the 0183 interface, since they cannot update the boat's position continuously the way a loran can. SatNavs do not command the pilot with heading-to-steer instructions, but rather with heading-to-waypoint, which nets a different track than the one plotted on your chart. SatNav is best for truly accurate fixes along the way, but loran is better at keeping you on course between fixes.

Actuators

After all the processing, modifying, and overriding done by the control unit, the result is an amplified signal that powers the drive motor actuating the rudder. The actuator may turn the steering wheel or tiller in the case of a cockpit-mounted pilot; the rudder cables, quadrant or steering rack in an underdeck system; or a hydraulic pump in a hydraulic steering system.

To adapt to the countless types of possible installations, there are three basic actuator configurations—linear, rotary, and hydraulic. Manufacturers offer one, two or all three types, depending on the extent of their product line. Probably all tiller-activating pilots have screw-type linear drives in which a servo motor (a small, high-torque motor often used to move aircraft control surfaces) turns

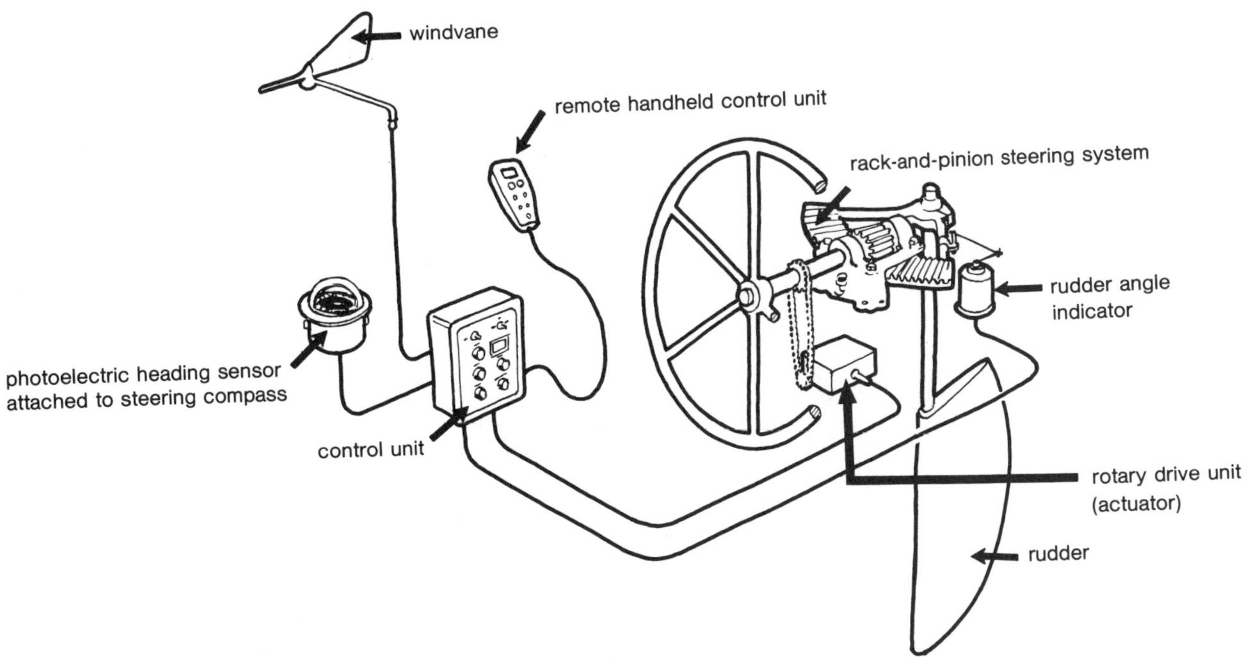

An autopilot for a large sailboat includes the components shown here. The steering system is a rack-and-pinion type, with the rotary drive unit actuating it through a chain to the steering wheel. A linear drive unit would most likely drive the steering rack itself, or a second quadrant just below it. The autopilot system shown here costs between $2,000 and $5,000, depending on options.

a long threaded shaft. Riding the shaft is a collar to which the tiller is attached. Many linear-drive units for belowdeck steering are no different, and usually are attached to a small dedicated tiller fitted to the rudderpost.

Rotary-drive actuators appear to be offered by more manufacturers, and are more likely to be seen on powerboats than sailboats. They can usually be mounted in more places than can the linear type, since they don't have the long throw of a screw shaft to accommodate, and don't have to be positioned next to the rudder post. Installation may be more complicated, though, requiring a chain to be spliced into the steering cables, the addition of a new sheave or two, and perhaps more structural modifications for mounting the new hardware. It is possible to mount a rotary drive unit so as to mimic a linear drive, by fitting a toothed rack to the quadrant and a pinion gear on the drive shaft. Cetrek's and Unipas's rack drive options, and Edson Corporation's curved rack segments for mounting on the rudder quadrant, are examples of this layout.

The least common in sailboat autopilot systems, though often the most sensible choice for powerboats, is the hydraulic actuator, in which the motor drives a pump tied into one of the boat's hydraulic steering lines. Few sailboats have hydraulic manual steering due to the lack of feel inherent in the system, but this doesn't preclude the addition of a hydraulic system for the autopilot alone. In a powerboat with anywhere from two to four steering stations (pilothouse, flying bridge, tuna tower, and cockpit in a big sportfisherman, for instance), hydraulics are preferable to a forest of cables and sheaves. There are plenty of hydraulic actuators available, since hydraulics are common aboard fishing and work boats.

A hydraulic system has the advantage of flexibility of installation. Fluid lines can be routed around bulkheads and previously fitted equipment. The

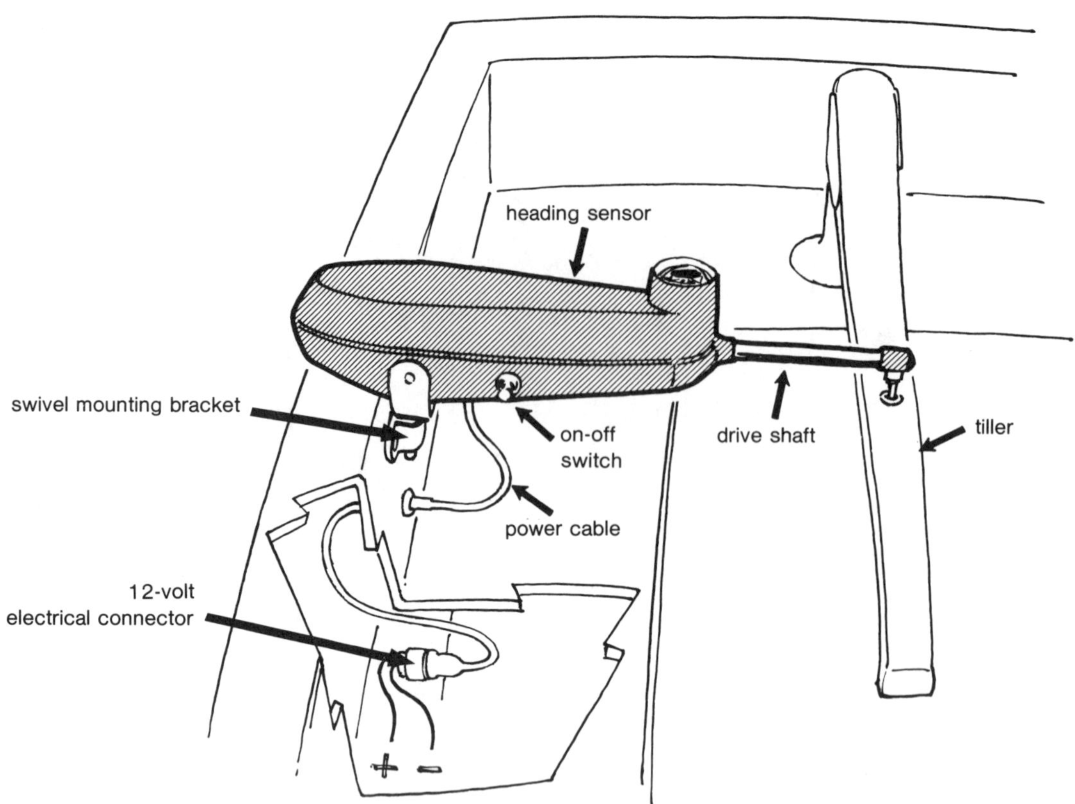

A beautifully simple setup is a tiller-driving autopilot that can be rigged or taken below in a minute or less. Some models come with the heading sensor separate from the drive unit, so it can be mounted below out of the weather. Most portables range between $500 and $800.

electric pump doesn't have to be mounted right with the cylinder at the quadrant. Boats with twin rudders can have one pump driving two compact cylinders. The autopilot system also serves as a complete second steering system, separate from the cable steering gear. There are potential fluid leaks and bubbles to contend with, however.

How much power will you need? Drive units are cataloged along rough guidelines according to boat length, but hull shape, displacement, rudder efficiency and helm-balancing ability, as well as likely cruising conditions, all play a part. To get a good idea of a sailboat's requirements, sail the boat in the strongest conditions in which you are likely to use the pilot, and at the point of sail (usually a close reach) that develops the most pressure on the rudder. You can simulate heavier conditions by overtrimming sails and creating excessive weather helm.

For a cockpit-mounted wheel-steering system for either power or sail, you can then measure the approximate force needed to turn the rudder with a formula suggested by the Benmar Company. Hook a fish scale to a spoke at the wheel rim, and pull enough on it to just turn the wheel, keeping the scale tangential to the wheel rim. On a powerboat, try the procedure at different throttle speeds. Higher speed turns should yield higher helm loads. Multiply the reading on the scale by the wheel radius. Multiply again by the number of turns lock-to-lock. The result is helm torque, in inch-pounds. This figure can be compared to torque figures on the spec sheets (to compare with manufacturer's figures measured in foot-pounds, divide your boat's number by 12).

For a tiller pilot, hook a fish scale on the tiller where you expect the pilot shaft to be mounted, given the throw of the tiller pilot's drive shaft. Measure the force in pounds needed to pull the tiller. Compare this number with the power ratings in the manufacturer's spec sheets.

This exercise is only valid if you plan to hook up the pilot at the point where you take the measurement (or if you are using Benmar's equipment). Gearing and friction in the manual system can mislead you as to how much power your autopilot will need at the rudder, where it will be mounted. The myriad considerations of an underdeck system really call for the help of a professional installer.

As important to the sailor as rudder turning power is power drain. How many amperes will the pilot suck out of your batteries while steering the boat? The answer will vary once again according to boat design, pilot installation, and sea conditions. While on standby, the pilot should draw a negligible amount of power, just enough to sense heading or wind direction. A typical draw for a pilot on course and not driving the actuator is 0.10 amps. Under "normal cruise conditions," which few manufacturers define, tiller-drive pilots draw from 0.20 amps to 0.33 amps. Momentary peak loads might call for three to five times as much power, but in such conditions you will be shortening sail and re-trimming to reduce the steering effort anyway. Makers of permanently mounted pilots list a wide range of power drain figures, from as little as that required by the cockpit-mounted models, to 10 amps or more while on "cruise." Some drive motors are rated much higher, but the rudder may well be hard over and stalled before the limit is reached. In any case, quoted power drain figures are only of value in comparing spec sheets when you know the numbers are obtained in the same way. How much power the system will use on your own boat is best estimated by the dealer or manufacturer, who has installed his pilot on similar boats and tested it with an ammeter. Be sure you know your batteries' capacity when you talk to the dealer.

Installation

Installing any pilot other than a tiller-steering model may involve more work than you think, especially if you have the manual steering system already in place. The simplest mechanical tie-in would be a rotary-drive actuator turning your boat's steering wheel shaft directly. Often there isn't room under the dash or bridgedeck of a powerboat for a big drive unit. Two manufacturers, King Marine and First Mate, make compact dashboard models for small motorboats. Self-contained like sailboat tiller models, they slip over the steering column and bolt to the dash. The steering wheel goes over the pilot. Some dashboard instruments may have to be repositioned to accommodate the pilot box. Benmar makes a similarly sized unit that goes under the dash rather than on top of it.

Sailboats with wheel steering do not necessarily have expensive underdeck autopilot systems. Among the most popular wheel steerers are Autohelm's 3000, Navico's WP-4000, CPT's Autopilot, and Tillermaster's new Wheelmaster, all of which feature motors that mount easily on the steering pedestal base and drive a detachable sheave on the wheel with a belt or chain. Using the mechanical advantage in the ship's steering system, these units can be smaller than the powerful underdeck drives yet still control surprisingly large boats. The winner of the first BOC round-the-world singlehander race carried an Autohelm 3000 on his 54-foot aluminum cutter. His only problem was that he used up a lot of rubber drive belts. For day-to-day cruising, these pilots are recommended for sailboats up to 40 feet.

More complicated, but more weatherproof, and in any event necessary for larger boats, is an underdeck system with the drive unit next to the rudderpost, driving the steering quadrant via linear drive shaft and tiller or through gears of chain via rotary drive. The layout of the boat's structure, the position of the rudderpost, its angle (vertical, or raked forward or aft) and the type of mechanical components used (cable and sheaves, push-pull flexible cables, rack and pinion gear, or even worm steering) determine whether a given unit will fit. This is where a professional installer is invaluable. He has seen many installations, knows what various manufacturers have to offer, and can help you select the one that delivers enough power and fits the space with a minimum of parts and labor.

Hydraulic installations in boats that already have hydraulic steering may be the easiest. The autopilot's drive pump can go almost anywhere that's handy, since you can run the hydraulic fluid tubes from it to a junction with the main lines.

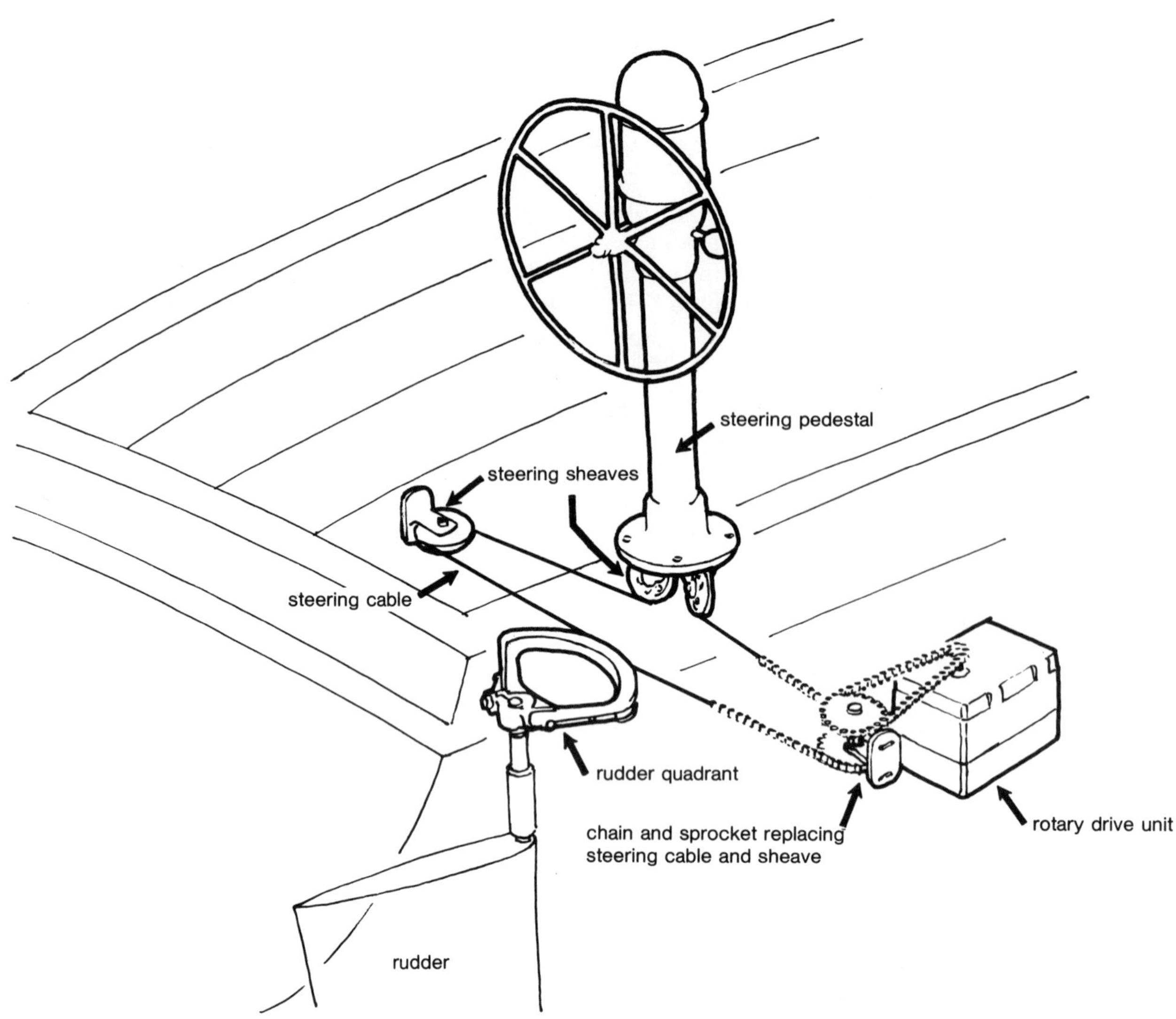

A common system for pedestal steering includes a rotary drive and a chain substituted for a section of one cable.

Alternatively, the autopilot system can be designed as a fully redundant backup to the manual steering. Naturally, you'll pay more for that safety.

A hydraulic autopilot system can coexist with mechanical manual steering, too. You do lose a certain amount of helm feel due to the damping effect of the fluid in the lines between hydraulic pump and rudder ram. The shorter the lines, the better. Still, such a system offers much more feel than an all-hydraulic system. Wagner and Cetrek both claim their hydraulic systems have finger-tip sensitivity. You'll have to try one out on a boat similar to yours to decide if they're right.

Your boat may well need structural reinforcement where the drive unit is to be located, as well as a mounting stand or bracket. Connections and the surrounding structure must be able to withstand as much power as the autopilot drive can apply to the rudder. Be prepared to modify your existing steering system to accommodate the new components. You may have to reroute cables so the pilot drive chain can be spliced into them. You may need a new steering quadrant that can carry both manual and autopilot cables, or a second quadrant facing aft. (Edson's catalog shows a number of different installations that can give you ideas.) A clutch may have to be installed in the manual system to deactivate the steering wheel and keep fingers from getting stuck between an unstoppable spoke and an unyielding steering console.

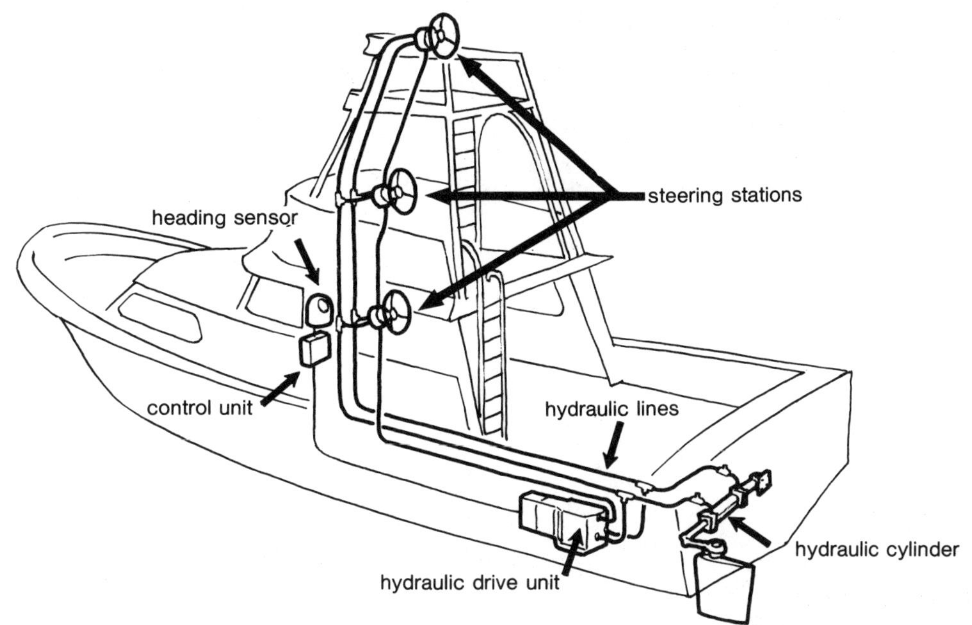

Hydraulic autopilot installations are quite flexible. The drive unit need not be at the rudderpost. Clutches disengage steering wheels from the system. Hydraulic pilot gear runs $2,500 and up.

Potential Problems

Once the system is installed, what kind of problems might be expected to crop up? First are those stemming from the installation itself. There may be binding of the moving parts under stress that would not show up at dockside. This will cause an excessive power drain, and mechanical wear. There may be magnetic interference problems. Even a well-installed autopilot can't be expected to work unaffected by noisy electrical systems nearby. Radio, navigation gear, stereo, instruments, and ships' systems such as the generator, bilge pump, battery charger, or even wires carrying current to light fixtures, can cause the compass to deviate from magnetic north. So can metal objects left too close to the heading sensor. A thorough trial period during which the pilot is directed on cardinal and intercardinal courses while each of the electrical systems is turned on and off is the surest way to avoid surprises later. Once again, this is where the experience of the professional installer pays off.

The other major worry concerning an autopilot's reliability is its water resistance. Cockpit-mounted pilots are most vulnerable to spray and boarding seas. With membrane keypads and good O-rings they can be well sealed, except where the drive shaft exits the case. This end can be better protected by adding a cloth, leather, or rubber bellows over the drive shaft. This won't keep all the water out, but it will help.

A more dramatic solution is to install your tiller-steering cockpit pilot underdeck. This may seem to contradict the reason for buying such a pilot in the first place, but many boats have the rudder post accessible through a cockpit locker or lazarette, so rigging or stowing the pilot would not be more trouble than it's worth. A separate tiller underdeck would have to be installed.

Pay careful attention to construction details prior to buying a cockpit-mounted pilot. Consider, too, the number of years the pilot you're looking at has been marketed. Five years should be enough time for the manufacturer to find out what works and what needs upgrading.

Out of sight must not be out of mind when installing a belowdecks system. Components should be in well-ventilated but dry quarters. External gear, such as plug outlets for a handheld remote control unit, must be regularly cleaned. Don't expect to be able to fix all the failed components yourself, either. With the shift to integrated circuits has come Mr. Fixit with his black briefcase full of miniature tools and spare processor boards.

Autopilots are a real boon to mariners who can't or shouldn't have to stay behind the wheel the entire time at sea. Just remember that the best autopilot is no match for the human hand on the helm when judgments have to be made. The autopilot can follow orders just fine, but it can't see for you or think for you. You're the captain, and you're responsible.

15

Performance Instruments

by Freeman Pittman

As with depthsounders and some of the navigational gear described elsewhere in this book, instruments —particularly sailing performance instruments— are no longer a case of one function, one dial. The range of capabilities available from a variety of equipment is so wide as to be impossible for most of us to master. But there is a cardinal rule to follow while voyaging through the world of time, speed, distance, heading, and angle: Don't buy it if you don't really need it.

Your choice in types of systems is growing faster than most people's understanding of them. There are the familiar single-function speed, depth, wind speed and wind angle, temperature, and heading instruments. There are systems that combine some of these functions to achieve a new set of numbers, called true wind and velocity made good (VMG). There are also some extremely complex systems that synthesize a great deal of information from dozens of sensors, and integrate not only all the above inputs, but also information from navigation electronics and from portable computers. It's sometimes hard to distinguish the subtle variations between instruments of one category and another. How much number-crunching power will you need in the system you choose? Do you want speed, time, and heading only for your dead reck-

oning calculations? Do you want that information to be available should you want to use it to compute speed made good to windward? Do you want a system that has potential far beyond what you think you need right now, or do you want a modular system that grows only as fast as your needs? Getting familiar with the terms and options should help you make the right decisions.

Speedometers for Power and Sailboats

Along with a depthsounder (see Chapter 2), the most basic instrument you're likely to need is the speedometer, or log unit. Of the two functions it displays—speed and distance—the latter is the more important. You can convert speed to distance and vice versa, but for navigational purposes, it's knowing how far you've traveled that's essential to knowing where you are or ought to be. How fast you were going along the way is secondary. Nevertheless, many speed-logs offer speed, distance, and time functions, allowing you to make your computations any way you choose.

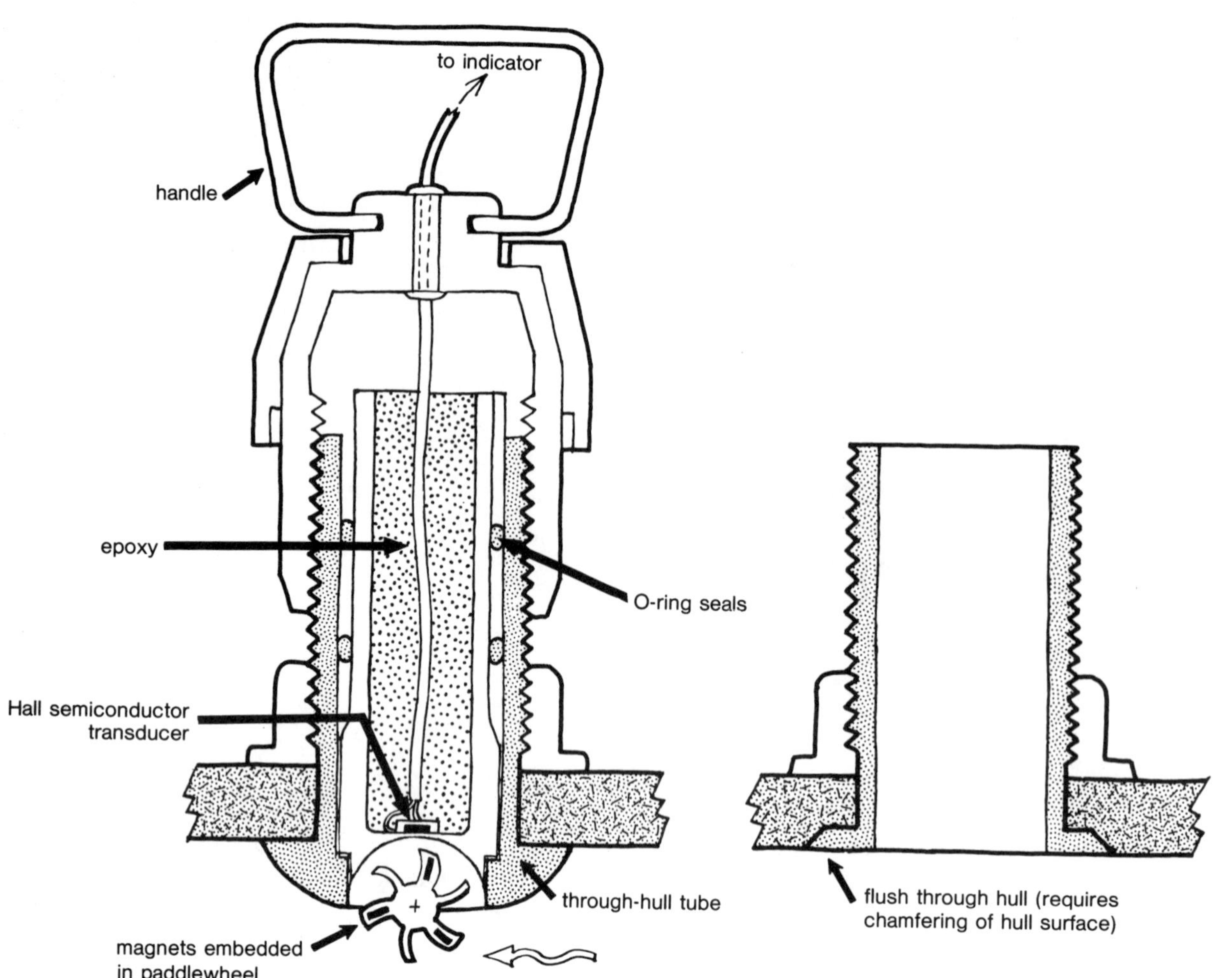

A paddlewheel boat speed transducer in its through-hull tube (left) and a flush-mounted through hull (right). Mounting the paddlewheel at the end of a protruding through hull can lead to some distortion of water flow, but the instrument's calibration will overcome most inaccuracies. Most through-hull units nowadays are plastic and must be mounted with compatible adhesives. They are also more vulnerable to damage than bronze tubes, but they present a small target. The boat speed indicator, paddlewheel, and through hull plus 50 feet or so of cable can cost as little as $200 or as much as $800.

Here are the readouts for a low-priced digital speed-log that is a features-for-the-dollar standard, the Aquameter 660 Speed/Log ($299.95):

- Choice of nautical or statute miles or kilometers
- Speed
- Average speed over trip
- Trip and sum (cumulative) logged mileage to 9,999
- Trend in speed (acceleration or deceleration)
- Elapsed time up to 99:59 hours
- Countdown to race start or any other event

Obviously, you can get much more for your money than just speed and distance.

The guts of the unit may differ according to manufacturer, but almost all modern speedos share certain features. The through-hull unit that generates the electrical signals used to calculate speed is normally a small paddlewheel impeller in a cylindrical housing. An amplifier sends the signal through a shielded cable to the display unit, where it is timed against a crystal oscillator, averaged, and displayed on a digital liquid crystal display (LCD) or on an analog (pointer-type) dial. The log portion of the display unit may be a separate mechanical readout like your car's odometer, or it may be a digital display that replaces the speed readout when you flick the switch.

Though most are similar in form and operation, transducers are the weak link in the system. There are a number of reasons why they can't be as accurate as they ought to be to keep up with the other electronics aboard, and there are several ways they can fail altogether. Many are the solutions that have been tried, but the vast majority of speed-logs employ the reasonably trustworthy and surely simple paddlewheel.

The plastic paddlewheel rotates on an athwartships axle. Molded into the paddlewheel, or glued to it, are magnets, and inside the through-hull unit just above the impeller is a sensor—an AC current generating coil, Hall-effect semiconductor, or electrical reed switch—as part of a circuit carrying a small but constant voltage. As the magnets in the impeller rotate past the sensor, they generate a pulse that is amplified and passed on to the indicating unit for processing. One way to understand what's going on is to envision the paddlewheel rolling its way along a nautical mile. An impeller with a 3-inch circumference will make four revolutions per foot, or 24,320 in a 6,080-foot nautical mile. Two magnets on the paddlewheel will result in 48,640 pulses per mile, assuming an ideally efficient rotation through the medium. Of course slippage must be expected, so speed-logs have rather wide calibration ranges, from 20 to 40 percent higher or lower than the displayed figure.

The above explanation is really more valid for the old-fashioned mechanical taffrail log, the spinner on the end of a rope dropped over the stern. In fact, electronic log manufacturers are moving quickly toward microprocessing, having already gone to solid-state components several years ago. Datamarine's Corinthian speedo, for instance, treats the transducer signal as an electric frequency, with about 4 cycles per second (4 hertz) equalling 1 knot. (To be sure they got it right, Datamarine engineers calibrated the transducer in a U.S. Navy flow chamber.)

Using the impeller's motion to generate its own signal without a current applied from ship's batteries is quite possible, and you can buy an inexpensive speedometer that does just that. The EMS U25K and SR Instruments' KL-5 are good examples. The impeller and through-hull operate as a

A typical indicator for boat speed and distance. Dimensions: 5 inches diameter over the bezel; 4 inches over the barrel, which protrudes through the bulkhead about 3 inches.

small generator to power an analog speed display. There is no log function, and 12-volt power is still required for night lighting, but for about $150 list, you do get basic performance recording.

With any paddlewheel, there are limitations on accuracy. In the case of a sailboat beating to windward, the boat's leeway combines with the asymmetric shape of the heeled hull to cause water to flow over the impeller at a different angle than while the boat is sailing upright. This crossflow makes for slower impeller speed. On the other hand, if the impeller is too close to the boat's keel, water around the keel's leading edge can accelerate flow over the transducer. Then there is the pitching and heaving of the hull, which make an impeller mounted too far forward vulnerable to breaking out of the water on wave crests. If the impeller is mounted in any spot on the hull that experiences frequent water turbulence, accuracy must suffer. And if it is on one side of the hull rather than on the centerline, it will probably read differently on one tack than on the other. This is a necessary compromise if the hull layup is too thick at the centerline for a through-hull installation. The obvious if somewhat expensive solution is to mount an impeller on either side of the hull, with a mercury or pendulum gravity switch to change automatically from one impeller to the other. In any case, the sailboat through-hull unit should always be mounted forward of amidships, preferably away from the keel or underwater fittings, so it will be in smooth water.

Displacement powerboat hulls don't create heeling, tacking, or leeway error (except when crabbing against a current), but they, too, need clean water flow over the transducer. Transducers on planing hulls should be mounted well aft, since the bow sections frequently operate in a lot of spray, and sometimes even in daylight.

One other factor often brought up when discussing the paddlewheel or other impeller types is the "boundary layer," the water closest to the hull. Due to the hull's natural surface friction, a certain amount of water tends to be dragged along with the hull as it passes through the medium. An impeller operating within this layer will read low. How far from the hull the boundary layer extends depends on speed, the shape of the hull, and location on the hull, but if a sailboat speedometer impeller is mounted correctly, it should protrude far enough into clean water to be accurate through most of the boat's speed range. Powerboats, with their much wider speed ranges, can expect greater accuracy problems at either end of the scale. At zero to 3 knots this is not a concern, but at 40 to 50 knots it is.

Manufacturers recognize the aforementioned factors and have ways to deal with most of them. Most speedometers have either a calibration knob on the back of the box or a programming button on the front that lets you change the displayed speed for a given frequency of impeller rotations. To find out what the right frequency ratio is, you run a measured course and compare it with the distance logged on your speedo. Use landmarks for course markers, since buoys can drift, and make your run when the tide is slack or the current is steady. Steer your boat on a straight course at a constant speed, as near as possible to its cruising speed under power or sail. Mark the measured distance after running the course, and then run the reciprocal

Aquameter's 660 digital speed/log indicator performs several functions in several modes quickly and efficiently. It exemplifies the current trend in instrumentation toward offering more information in a smaller package.

course. Add the two distances and divide by two. Compare with the charted distance and adjust your speedometer, according to the manufacturer's manual, to resolve the difference. Errors of 2 percent or more can be eliminated in this way.

What can't be adjusted for is an impeller fouled by marine growth or floating debris. Mount the impeller where it's easy to reach for inspection, and check it occasionally. Clean as needed. Some units are painted with antifouling paint at the factory. This can be renewed, but check with the manufacturer as to what paint is compatible with the plastic, and be careful not to gum up the works.

For all its potential problems, the paddlewheel and its close cousin the propeller-type impeller work pretty well. They're also relatively economical. But inventive minds are always trying to improve on the things that aren't yet perfect, and the area they're focusing on is solid-state transducers.

Right now the only such system that looks like a model for imitators is Brookes & Gatehouse's Sonic Speed. Developed from submarine acoustics research in England, the system consists of two small ultrasonic transducers similar to miniature depthsounder probes. They are mounted 12 to 36 inches apart on the hull, or on the hull and keel, facing each other along a fore-and-aft axis. There must be clear water between them, a condition easily satisfied on most of today's sailboats featuring a

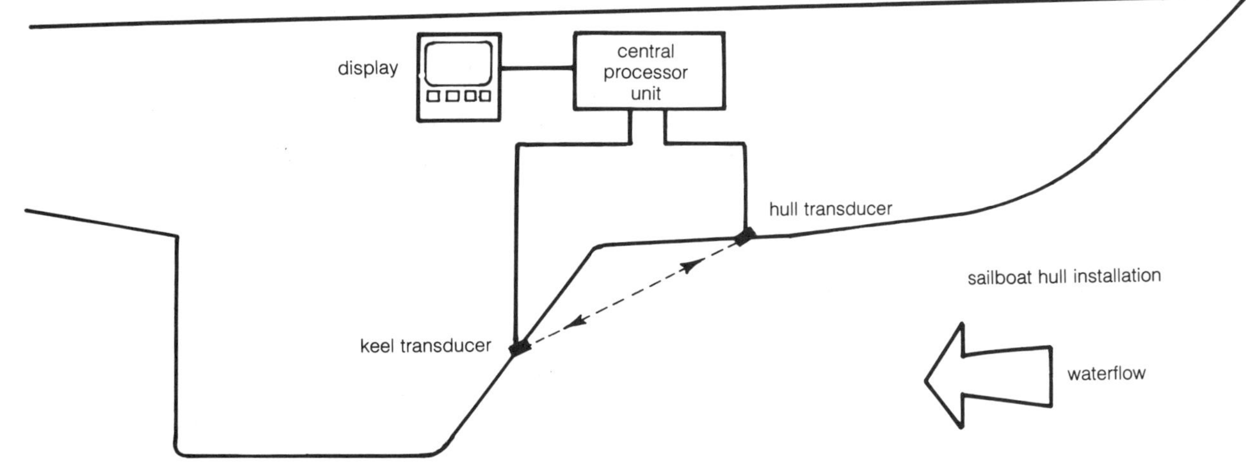

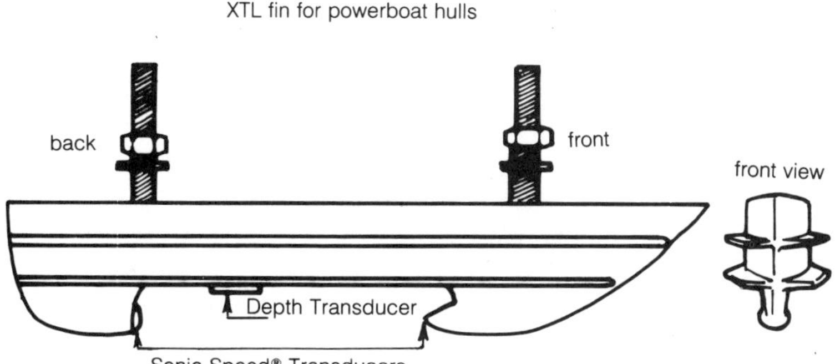

The Brookes & Gatehouse Sonic Speed water speed transducer works on the principle that water flow speeds up sound signals moving in the same direction, while slowing down signals moving against it. The processor unit calculates the difference in speeds of the ultrasonic signals passing between two transducers, and converts that to a displayed boat speed that is very accurate in all speed ranges.

cut-away forefoot. On many powerboats, the units can be fitted to the hull and skeg. B&G has produced a plastic fin enclosing both speed transducers and a depth transducer for easy mounting with a couple of bolts through the hull.

Each speed transducer sends out a 171 kHz signal. The signal sent aft is hurried along by the water flowing with it, while the one sent forward is slowed. The speedometer's microprocessor subtracts the slower signal from the faster and comes up with twice the boat's speed, which is halved and displayed. The device does away with boundary layer problems, moving parts, or water piling up inside the through-hull and slowing an impeller. Fouling from marine growth must still be prevented, and the transducers cannot be removed from inside the boat. Also, turbulence can affect the readings, just as in depthsounders, so the underwater units must be sited in clean water. There are two major limitations on Sonic Speed, though: it only works with B&G hardware, and the basic system adds at least $1,000 to the cost. Nonetheless, the next couple of years will see this fascinating mechanism proven and copied.

The coaxial cable carrying the transducer's signal can pick up interference from other electronics if it's cut or damaged. Don't cut it to length without first reading the manufacturer's instructions—they may warn you not to cut it at all, but simply tape the excess in a coil and secure it. Good-quality instruments have the transducer and cable end potted in epoxy, from which the cable runs about 10 feet to a connector. The connector enables you to remove the through-hull fitting if needed without having to cut or pull out the main run of cable. There should be a connector at the indicator unit end of the wire, too.

Racing sailors, along with cruisers of both sail and powerboats, will appreciate a timer function in their speed-log. If it has a countdown function, it can be used for racing starts. As a chronometer for navigation, it should be able to give elapsed time, and should be easy to synchronize with radio time ticks.

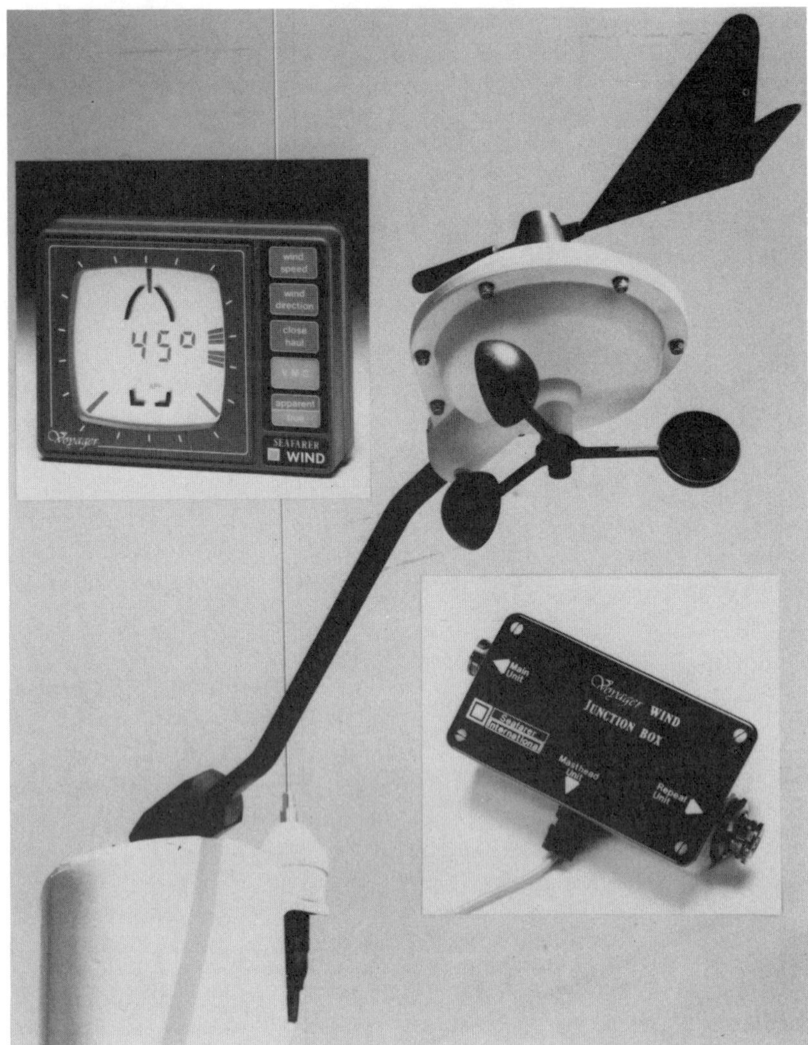

Despite a rather bulky masthead unit that houses an unusual photo-optical wind angle sensor, the Seafarer wind speed and angle instrument offers some advanced features. The large liquid crystal display has graphic as well as numeric representation of wind angle—both apparent, and true. When interfaced with a Seafarer speedometer to form a simple integrated system (see page 138), it also reads out speed made good.

Wind Instruments

While every mariner should learn to judge wind speed by how it affects the water's surface (the Beaufort scale for wind force), and be able to feel where the wind is coming from without looking, an electronic wind speed meter can measure puffs and lulls your eye and skin can't always distinguish. And while every sailboat should have a non-electronic masthead fly such as a Windex to show apparent wind angle, a wind angle indicator in the cockpit provides an excellent fine-scale guide to wind shifts and shows clearly which tack is favored upwind. At night, when your own senses are limited, wind instruments become more valuable. The wind speed and bearing you feel, which your boat's instruments measure in raw form, are called "apparent wind" functions. They combine the true wind speed and bearing with the boat's speed and heading. The resulting apparent output is very useful for gauging boat performance and weather trends, but if you're interested in keeping up with the competition, you may have to start looking at some more precise, computed wind data. When you decide to use this "true wind" and "velocity made good" data, you enter a world of sailing performance technology and philosophy perhaps unlike what you're used to. We'll show you a bit of this world later, but first let's look at the hardware used for all wind measurements.

Wind speed instruments (anemometers) have much in common with speed-logs. Instead of having a rotating impeller underwater, they have one aloft, in the form of three little black cups spinning

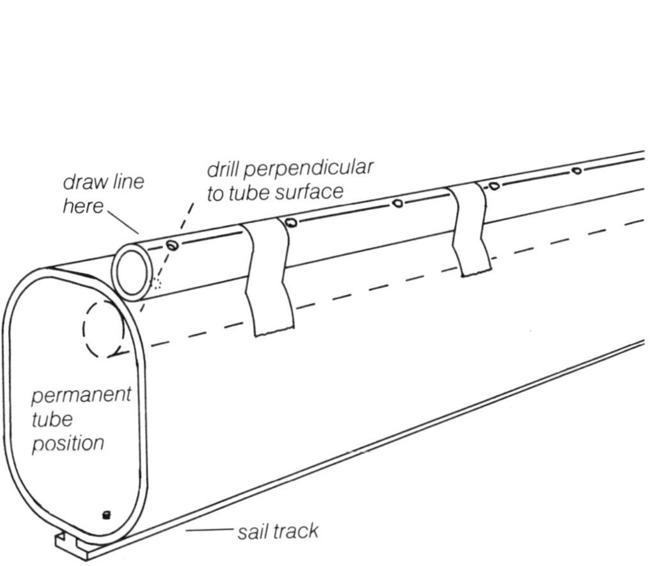

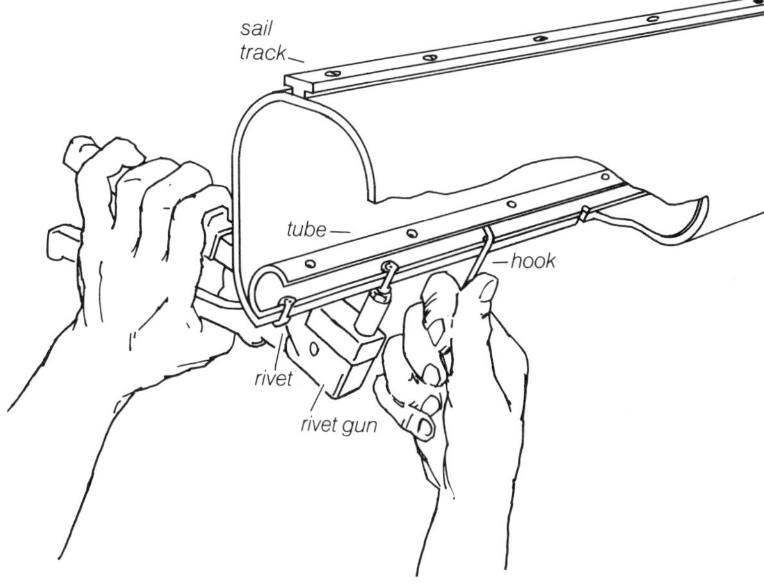

Installing a PVC conduit for electrical wires within the mast. **Left:** *Firmly tape the PVC tube to the outside of the mast just where it will be permanently on the inside. Draw a guideline for drilling on the tube 180 degrees from the line of contact between mast and tube. Drill perpendicularly through the tube and mast.* **Right:** *Use hooks to align and hold the tube for cutting exit wire holes and for riveting. Wiggle the hook in the hole next in line to the one being centered. Use the hook to hold the tube snugly against the mast, and rivet. (Courtesy* Sail *magazine)*

in the breeze, or, on Silva Instruments 4000 masthead unit, a propeller like an airplane's. The rotor creates electromagnetic pulses with the same hardware used in the water-speed through-hull. In addition to a Hall effect or generator sensor, a photocell may be used to count flashes from an LED as seen through gaps in a fitting that spins with the rotor. For most wind speed sensors on the market there is no calibration circuit, since there is no hull or boundary layer to affect fluid flow around the sensor.

The transducer is normally mounted on the end of a stalk facing forward from the masthead. This gets it out into clear air and away from antennas, halyards, and sails—especially the spinnaker—which might otherwise foul it. You might be surprised by the amount of aerodynamic interference other hardware aloft—including even the wind unit's own stalk—can develop. This is more likely to affect the wind angle vane than the anemometer, however. The masthead unit must be robust, but with very light moving parts, and it must be water- and corrosion-proof.

The masthead unit installation requires effort and care. Be sure the unit is securely positioned on the fore-and-aft axis and that it indeed extends forward from the mast, unless the manual specifically states that you can mount it facing aft over the mainsail. The wind speed unit won't care, but the wind angle indicator that usually comes with it may not operate properly unless facing forward.

If you are lucky, your mast will have a separate tube for running electrical wires, and if you are thoughtful, you will have a permanent messenger line inside the mast or tube to draw new cables from deck level to the masthead. Owners of boats without a dedicated tube may want to add a PVC conduit on the inside of the mast, as shown in the accompanying drawings. The cable can also be secured to the outside of the mast, but that exposes the coax to the elements, and ensures a short life for it. If possible, have the cable exit the mast below the deck but well above the bilge. When planning this arrangement, be sure the cable and its connectors will pass easily through the mast partners when it's time to pull the mast out of the boat.

Calibrating a wind speed sensor is only possible with some of the integrated instrument systems. It can be done with a handheld anemometer you know to be accurate. You have to go aloft and hold

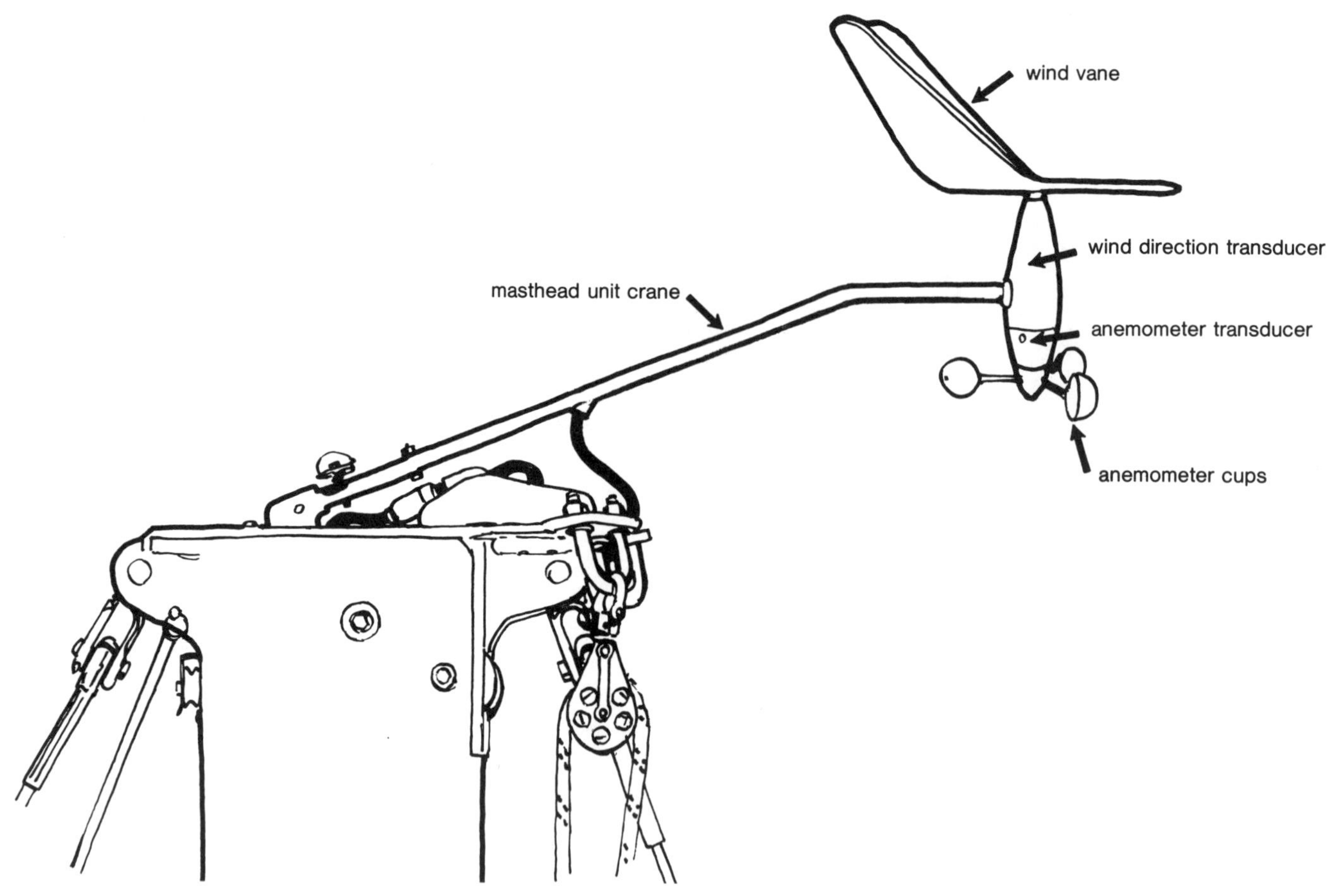

A masthead combination wind speed-wind angle unit extends forward of the spar into clear air. Components are lightweight aluminum for low inertia, but they're easily damaged. Connections must be well sealed here or you'll spend a lot of time in the bosun's chair.

it near the masthead unit. Do it on a day with variable winds, to get a range of wind speeds for comparison. Someone on deck records the readings from the masthead unit as you note the anemometer speeds.

Calibrating a wind angle unit may be trickier. Assuming the display is adjustable, sail the boat at a steady speed on both tacks. Note the corresponding wind angles and adjust the display by half the offset. You'll probably need several tries to get a reasonable estimate.

The wind angle vane rotates on a precision bearing, and it should balance perfectly. You can be sure it does by holding the unit on its side and watching to see if either the nose or the tail of the vane drops. Often the nosepiece is adustable for weight distribution, or you can add small weights.

The vane's shape varies with the manufacturer's aerodynamic theory (or lack thereof). There's no way of telling in the store whether the vane you're inspecting is going to fly well on your boat. Ockam Instruments, which makes as sophisticated an integrated system as you can buy, doesn't make its own sensors and recommends Brookes & Gatehouse or Signet to users.

To generate electrical signals, the windvane likely has a metal wiper or two fitted to its shaft inside a housing just below the vane. The wipers sweep a potentiometer coil wrapped around the shaft. Varying resistance around the coil results in different voltages for the various angles sent to the display unit on deck, and these are converted to specific readouts.

The potentiometer in such a masthead unit is

likely to wear out after five seasons. From the moment it's installed it is working, and salt air eventually finds its way in to corrode the contacts. Some manufacturers, therefore, prefer a non-contact circuit that uses a magnet on the vane shaft to change the voltage in a coil surrounding the shaft. More expensive, but it has several advantages. One, the friction from the contacts is eliminated. Two, the case can be filled with a light oil to damp the vane's movement. Three, the system allows full 360-degree readings. With most potentiometers there's a gap of 10 or 20 degrees to either side of dead ahead that simply doesn't register. This may manifest itself as a jump of the pointer across the dial from port to starboard, or as a dial whose degree marks aren't where they should be (see wind indicator display drawing). Check this out in the store before you buy.

On the other hand, if the needle moves from one angle to the next in discrete steps like the second hand on a clock, it probably has a stepper motor in the display case to track the signals from the masthead unit. That's what Datamarine and Coastal Navigator put in their products, and it's a precise means of repeating what's happening at the masthead.

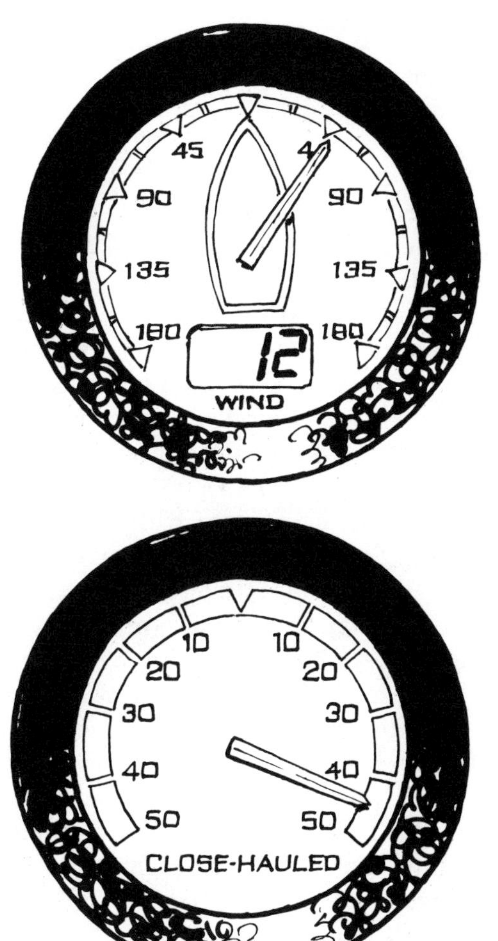

Standard readouts for apparent wind angle on the 360-degree scale (above) and on the expanded close-hauled scale covering only the first 60 degress on either side of head-to-wind (below). The close-hauled option is helpful to racers. The angle displayed is relative to the boat's heading. Note that the bearings on the 360-degree scale are not placed as they would be on a compass card. This indicates the masthead unit does not record wind angles all the way around the scale. Wind speed is often combined in the same instrument, as it is here. The package (including sensors) costs $600 and up.

Display Unit Features

The display unit, whether for boat speed, wind, water depth, heading, or other data, takes one of several shapes and forms. There are the traditional barrel-shaped models with a flange or bezel for mounting through a bulkhead. Other models are similar but with a square bezel. There are nearly flat units that mount on the bulkhead. These do not require cutting any major holes in the bulkhead for mounting; you need only drill holes big enough for screws and wires. There are wheelhouse units with trunnion brackets for the dash or overhead on powerboats or on sailboats with pilothouses. And there are display units that combine several instruments in one housing, either with separate displays or sharing a single readout. This usually doesn't allow for as much flexibility in installation as do individual instruments, but it sometimes makes for a lower package price.

Do you want analog readouts or digital? The choice is personal, but digital displays can give more discrete readings. Many digital speedometers read out in tenths of a knot, and some even in hundredths. That implies greater absolute accuracy than the system is capable of producing, but for the racer interested in seeing trends displayed, it's a worthwhile feature. Though a given readout may not be precise, the numbers are consistent, and accurately signify the magnitude of speed changes.

The only instrument for which an analog display is preferable is wind angle, although here again a finer reading is had with a digital display. You just have to look up at the masthead fly to remind yourself what the numeric readout means.

Every point made in the chapter on depthsounders about LCD versus LED holds true for speedometers and wind instruments. Actually, though, it's hard to find any digitals that don't have LCD readouts. In any LCD, contrast and clarity in bright sun depend very much on the quality and cost of the liquid crystal specified for the instrument. Also pay attention to how wide a viewing angle the display has. LCDs are not noted for showing much unless you're looking at them nearly head on. If you steer from the windward or leeward rail, as do many sailors, the instruments will have to be mounted well ahead of the helm or in one of the swiveling instrument pods you can get for steering pedestals from Edson or Yacht Specialties.

Instrument lighting is something you don't often get to judge in advance, but do try. Often the lighting is too bright, and you can't look from the instruments to the sails or horizon at night without waiting for your eyes to adjust. A brightness adjustment on the instrument is a nice feature.

The other aspect of readability you don't get any indication of in advance is resistance to fogging. If the instrument has a sealed double-layer lens or is hermetically sealed altogether, you should have no problems. Not many units have these features, however.

Integrated Systems

So much for the basic wind and water indicators. But we have other uses for their outputs. From them can be derived what are called computed functions, data that are the result of combining wind and water inputs and adding a dash of trigonometry. The most important of these computed functions are velocity made good (VMG), true wind speed, and true wind direction.

True wind direction and speed, the values you'd measure if you were sitting atop a buoy rather than on a moving boat, if not critical are indeed worth having for the sake of safety and good seamanship. If you are sailing directly downwind with the apparent wind indicator at 15 knots and the speedometer at 8, you know the wind is blowing 23 knots true. That means you'd better have your heavy headsails ready to hoist and a reef or two in the mainsail before turning upwind. This calculation is easy to make dead downwind or directly into the wind, but at all other angles it takes some fancier thinking than most of us could do while steering and keeping watch. The basic equations are shown below.

True wind direction means you can select the right sails more easily before heading off on a reach. Will you be far enough off the wind to set a spinnaker, or must you keep the genoa up? Or perhaps the reacher and a staysail make the right combination. This judgment is not easy to make before you've actually turned onto the new course, and if you've guessed wrong, you must go through a needless sail change. True wind direction is a certain guide to use in conjunction with a polar chart or table of wind angles to match the sails in your inventory to the expected conditions.

Velocity made good tells you which course and speed is the most efficient for getting to a point directly upwind or downwind. Will pinching the boat higher and trimming the sails in tight result in a net gain to windward even if you're going noticeably slower on the boat speed meter? Or should you slack sheets a bit and foot off faster on a course not so close to the wind? VMG tells you this, as well as the best angle at which to reach broad off the wind, and more.

All these functions are useful, but they're impossible to solve in real time without the integrating power of a microprocessor to bring together wind speed, wind angle, boat speed, and for some functions, compass heading, on an ongoing basis. True wind is the key to the other computed functions. It must be found first; the others follow.

$$V_t = \sqrt{V_s{}^2 + V_a{}^2 - (2V_s \times V_a \times \cos D_a)}$$

$$D_t = D_a + \arcsin{(V_s \times \sin D_a / V_t)}$$

$$V_{mg} = V_s \times \cos D_t$$

where V_t = true wind velocity
V_s = boat velocity
V_a = apparent wind velocity
D_a = apparent wind angle
D_t = true wind direction
V_{mg} = boat velocity made good

The computer's memory supplies square root, arc sine, and cosine values, along with the simple math. To be really accurate, leeway must be accounted for in the true wind direction value, but that requires sensors not included in most computed function systems. V_t, D_t, and V_{mg} are the building blocks for a series of other functions. These include:

- Distance lost to weather on each tack;
- Speed or distance gained to leeward;
- Optimum theoretical boat speed based on polar speed curves of the boat's performance potential;
- Difference between actual speed and theoretical potential;
- Predicted apparent or true wind direction and speed for the next leg to be sailed;
- Optimum tacking angles upwind and jibing angles downwind;
- Distance made good to the next waypoint;
- V_{mc}, or target boat speed. This is a different way of perceiving performance, discussed later.

There are many more to choose from, and you can invent some of your own by programming a personal desktop or portable computer linked to the system through a standard interface such as RS-232. See the sidebar for some computer programs currently available.

Computed functions come to you plain, fancy, or very fancy. The choices are varied, but in the most general terms, instruments that produce these functions can be put into one of two groups: independent unit systems, and central processor unit (CPU) systems.

Independent Unit Systems

Independent units usually look and work like their basic one-function counterparts, but they have more sockets in the back and more buttons in the front. The sockets allow inputs from not only,

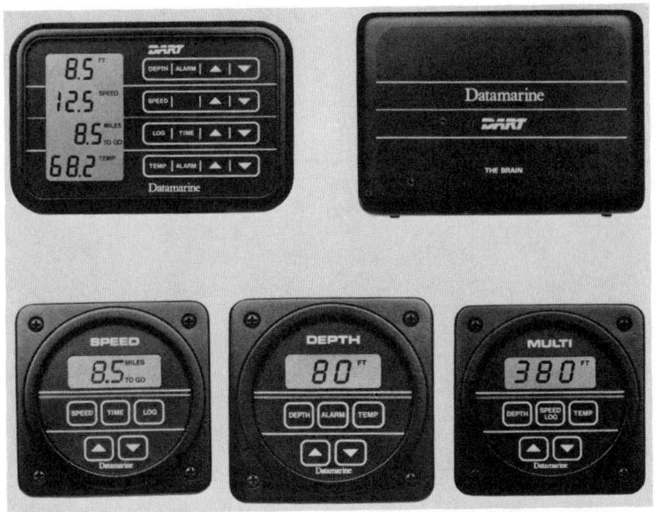

Datamarine's Dart line of instruments include speed/log, depth, and water temperature readouts in various combinations. Although a central processor box is used, and more than one readout can be called up on any display, this is still primarily a single-function system.

say boat speed, but also from wind speed and wind angle transducers. The buttons let you call up, say, three to ten times as much information as you can get on the display of a comparable single-function model. A trend in multifunction displays is toward larger LCDs with graphics that supplement the alphanumerics. The Silva 4000 wind indicator, for instance, shows you digitized pointers for both apparent and true wind direction, with wind speed numbers displayed below. It can show reference lines for tacking angles, and it flashes at you when you sail by the lee downwind. Most independent unit systems, however, do not give magnetic wind direction—the source of the wind relative to magnetic north. An electronic compass is needed to produce this readout. Instead, they give true wind angle relative to the boat's bow, which is zero degrees on the boat's wind indicator dial. Usually, it's the more expensive systems described later that incorporate a compass. In addition to Silva, independent unit systems include Robertson-Shipmate's RS series, IMI Stowe's Navigators, Signet's 1000, 1500, 2000 series, Navico's 200 series, and Seafarer's Voyager series.

Such "smart" instruments take a little while to learn, and before buying it would help to pick up a

Software For Integrated Systems

There is not much available on diskette to slip into your laptop computer, plug into your integrated system's RS-232 port, and make you a more competitive racer or a more informed cruiser. And some of what is on the market is very high-priced. But keep on the lookout, for race tactics, navigation, and weather programs are beginning to make an impression on the sport.

The most action, to be sure, is in the sailboat racing arena. The driving force has been 12-Meter match racing, and 1988 should see commercial versions of programs developed for the Twelves offered by Ockam and Durgan-Wake Associates. Right now, Ockam offers a line of programs called Ockamsoft. The simplest and least expensive is for the Epson HX 20 laptop computer, offering downwind jibing angles and next-leg apparent wind direction, among other functions. The next more sophisticated program is for IBM PC computers and clones. During a race it provides continuous recording of up to four user-selected functions. It gives optimum downwind course to steer, next leg data, and other piloting advice. And all the information it puts out goes into files that can be entered in a Lotus 1,2,3 spreadsheet program for exceptionally useful post-race analysis—where were we on the race course, where was the wind coming from at any given time, were we sailing to our target speeds, and so forth.

Compusail, by Durgan-Wake Associates, is a similar program that's been well tested with Ockam and Brookes & Gatehouse instruments. It, too, runs on IBM compatibles, most popular being the Data General DG One, and Zenith 181. Compusail can construct, modify, and save race courses, give your position by loran or dead reckoning, project polar curve data such as target boat speed (in case you don't have an instrument system that derives such outputs), give optimum course to steer while reaching (taking into account leeway), predict whether the boat must be jibed or tacked on this leg or the next, and present, analyze, and save boat performance and conditions data in various ways. It can modify polar data based on sea condition and sail configuration, and it can correct for inaccuracies in instrumentation. It displays and prints a range of charts and graphs, in single or multiple colors.

Brookes & Gatehouse, meantime, has a program for an Epson PX4 laptop. It's called Tacstar, and it's actually "firmware," software programmed into a chip that's added to the computer. That's a good feature in the marine environment, which is very unkind to traditional floppy disks, though it also limits the program's size, power, and flexibility. Tacstar works with B&G Hercules 290 or 390 systems. It provides dead reckoning navigation, though it works with loran-C as well. It monitors, analyzes, and updates the boat's performance numbers in the Hercules trial horse program, so unrealistic speed expectations can be easily modified. It displays a wind history in graphic form that can last from five minutes to 10 hours. It computes current and tides from loran input, or based on numbers input manually beforehand. There are other, more specific outputs, and the list is growing.

Of all the programs currently available, only one does what the casual observer thinks a computer is supposed to do on a boat—tell you which way to steer. Designed in France for the Apple MacIntosh computer, it is called MacSea. It is marketed in the U.S. by Ockam. MacSea integrates data from instruments, from a velocity prediction program, and from radio-telexed weather data, along with manual inputs for sail configuration and currents, and offers you a selection of courses for the next 24 hours, and how far and fast you'll go on each of those courses. It displays your track and projected course on a digital chart. It has already helped several French ocean racers make record crossings of the Atlantic, and the leaders in the most recent BOC singlehander round-the-world race all used it. Ockam thinks a bigger market lies in commercial and deep sea fishing, however, where an efficiently shaped course saves a lot of fuel dollars as well as time. Expect to pay up to $7,000 for a MacIntosh equipped with MacSea.

A host of peripheral programs relating to digital atlases, starfinding, and celestial navigation are available for everything from Sharp pocket computers to Tandy laptop computers and full-size PCs. Nothing is as well defined or as complete as the racing programs, however, and cruisers looking for MacSea-like programs will have to wait another year or two.

Race course mark coordinates, distances, and headings
for Gold Cup, Olympic, or offshore courses.
Predicted apparent wind angle and speed for next leg.
Predicted true wind angle and speed for next leg.
Target boat speed for prevailing apparent wind—now
and on next leg.
Recommended sail configuration for conditions and
course.
Predicted mode to next mark—tack, reach, or jibe.
Optimum heading to next mark.
Optimum heading on next jibe or tack.
Optimum reaching course—accounting for leeway.
Distance and ETA to various marks and waypoints.

GRAPHIC

True wind speed and direction fluctuations plotted
over time.
Laylines to mark as wind shifts.
Tacking efficiency plot—for comparing with other
watch.
V_{mg} plotted over time—also for comparison.
Current set and drift vectors.

*Some computed data available in current software
packages. These are only a few of the many functions
that can be programmed or bought as ready-to-go
software for computers such as the IBM PC, the PC
Convertible laptop, the Apple IIe, the Apple
MacIntosh, the Epson HX20, the Hewlett-Packard
HP-41, the HP-75, the Compaq, the Data General
One, and others. Consult the integrated system
manufacturer for a list of compatible software
and hardware.*

copy of the owner's manual and spec sheets for the
system you're interested in. Familiarizing yourself
with the functions and terminology makes understanding sales claims and descriptions easier.

CPU Systems

The central processor unit systems include all the
cutting-edge electronics for sailing. This is the stuff
carried aboard America's Cup 12 Meters, International Offshore Rule ocean racers, and record-
breaking transoceanic multihulls. A CPU system is
a bunch of transducers, sensors, and control units
that feed a central processor. It then spews out
information to system displays (none of which can

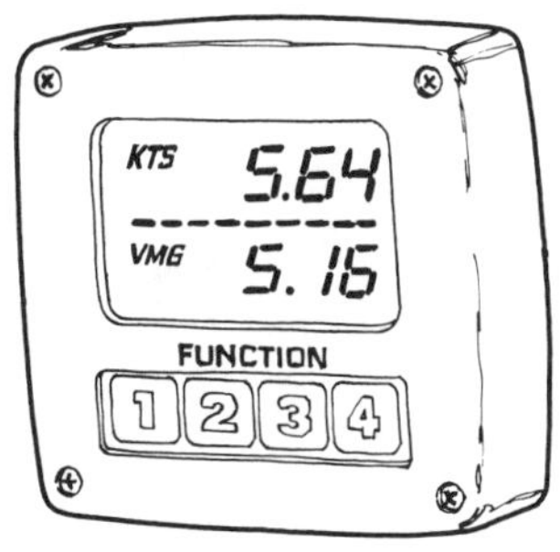

*Digital multifunction displays allow readings
from several sensors to be shown in one compact
box. This box may simply display the data from
the sensors, or it may use the data to calculate
additonal functions such as speed made good
(V_{mg}), true wind speed, or true wind direction. In
the latter case, the box is part of an integrated
system, and the prices for such systems range
from $800 up.*

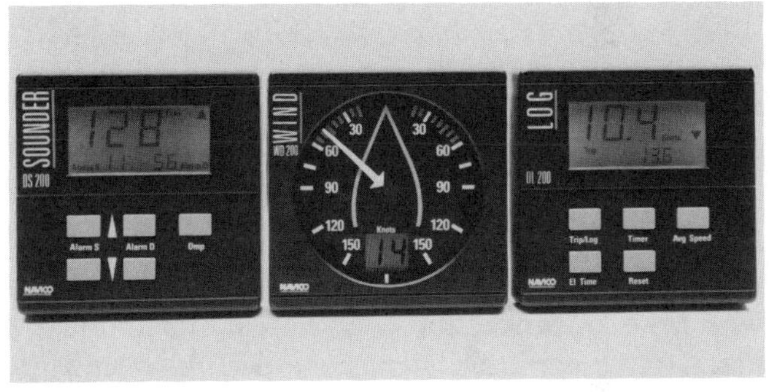

*Navico's 200 series independent unit instruments
use microprocessors to develop a limited number
of computed functions such as speed made good.
With watertight housings and a slim profile that
requires only fastener-sized holes drilled in the
bulkhead, instruments like these are fast replacing
traditional barrel-shaped analog indicators.*

function independently), a printer, or an outside
computer. (In the case of America's Cup boats,
every bit of data from every source is transferred by
radio telemetry to a chase boat and a shore station
for analysis.) An onboard laptop computer can be
programmed to display a whole range of special
functions (see table for examples). The CPU can

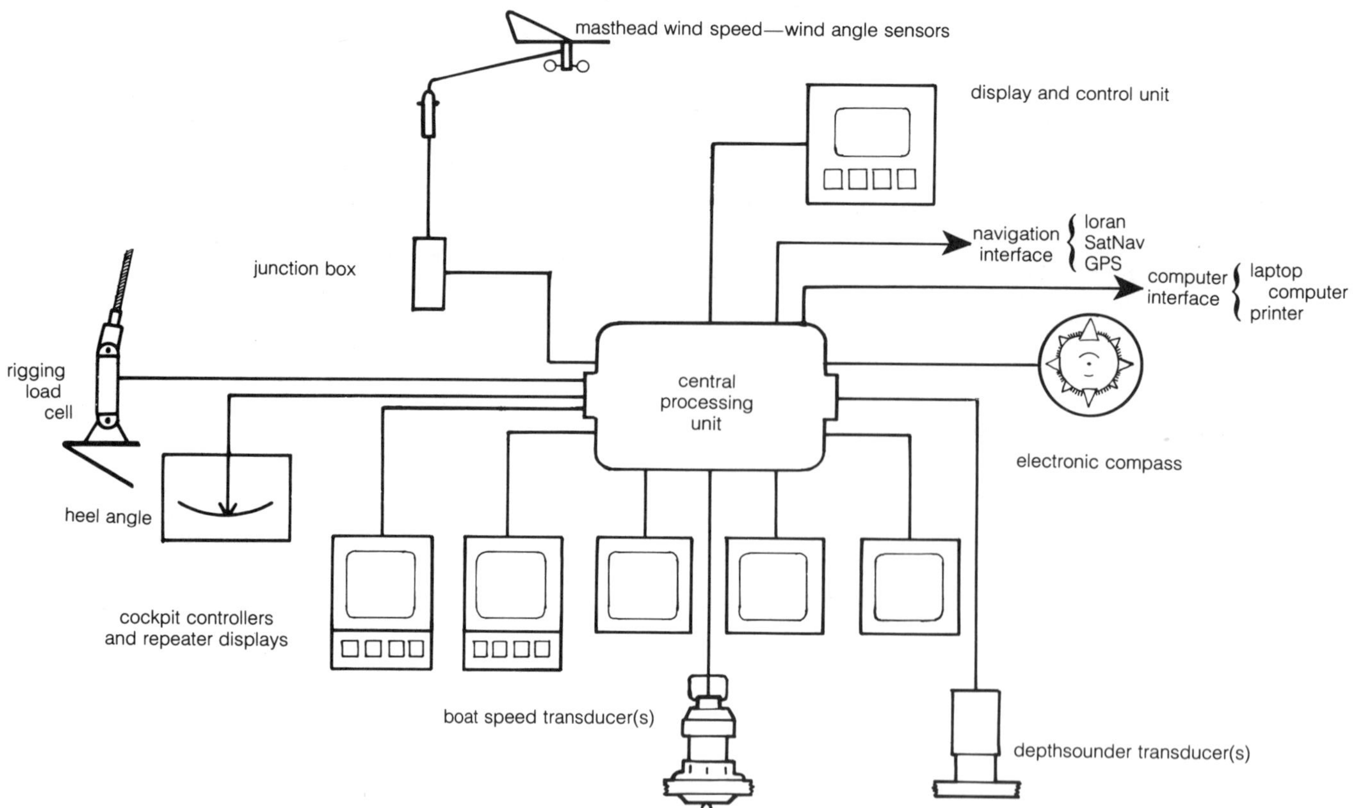

A central processor-based integrated system feeds inputs into the CPU from speed, wind, heading, depth, and heel sensors, and from outside units such as loran, SatNav, or a computer. The CPU then distributes the calculated data to the appropriate readouts in the nav station and cockpit.

also take inputs from a SatNav or loran receiver and display true position on the race course along with time, distance, and course to marks and laylines. It can compute current set and drift, or leeway. It has its own electronic compass, for which it memorizes compensations for deviation and variation, and it can display true or magnetic bearings and courses, along with their reciprocals.

It includes a host of calibrations and corrections for such things as heel angle and upwash on the wind angle readouts. B&G and Ockam CPUs can also be equipped with a velocity prediction program (VPP). Other systems, such as Atlantic Instruments, Datamarine Link, and Signet Smartpak, can, too, though they require a computer and software to do it. VPP is a table of performance targets in various wind and sea conditions based on your boat's design parameters and different sail combinations. You get a realistic view of how well you're doing relative to the optimum speed you can expect to achieve. The readout can be in knots, or in a percentage of the predicted speeds. It helps in tuning the boat—racer or cruiser—as well as during a race, when your competition is too far away to serve as a yardstick for judging the effects of your trim changes.

A few integrated systems do what CPU systems do without depending on a single central processor. They are "network" systems. Each of a network's components functions independently, but also communicates with the others to a much greater degree than the independent unit systems. The most advanced network system—Datamarine Link—uses a single coaxial cable to carry system power, as well as all the various dialogs, from component to component. Network systems offer a kind of high level redundancy for safety's sake, but they also provide owners with tremendous flexibility in system development.

Installing, calibrating, and learning to use an integrated system are not simple tasks, but you should take the time to do as much of the work as

A few of the components of a CPU-based integrated system are shown here: Danaplus Danavigate 7000 CPU (the black box upper right); the multifunction control and display (lower right); and sensors for speed, wind speed and direction, depth, and heading.

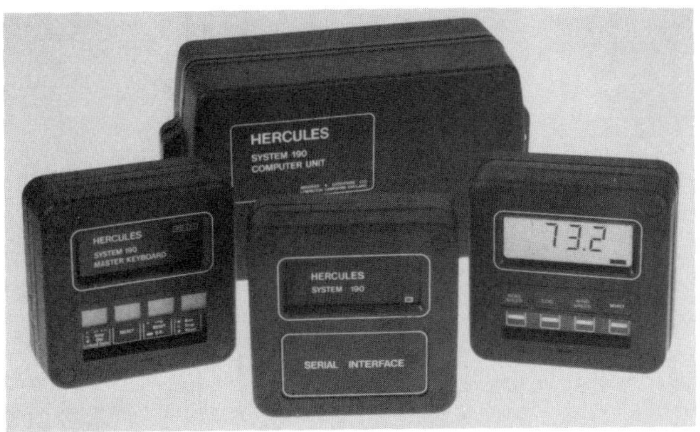

Brookes & Gatehouse offers a wide variety of readouts in its central processor—based integrated Hercules system. The CPU is the large box in the background; clockwise from it are a multifunction cockpit display, a serial interface box for tying in a portable computer, and a master control keyboard for the CPU in the nav station.

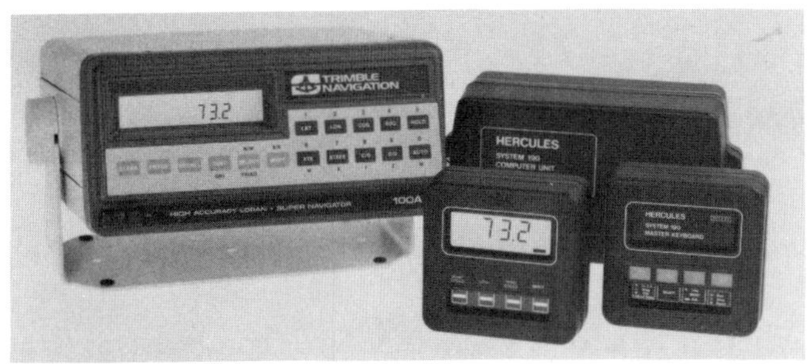

Brooks & Gatehouses's CPU-based Hercules integrated system can interface with a loran such as the Trimble shown here, a SatNav, or with personal computers via serial interface.

possible yourself so you can understand the system and use it to its potential. Unless it's a coax cable wiring job, though, a wiring expert may save you trouble by installing the components right the first time, with a minimum of resulting electrical interference. Most of the wiring connections in these systems are solderless, multiprong plugs and sockets that only go together one way. It's isolating the sensitive electronics from engine, radio, and motor noise, putting the electronic compass in the best possible position, and compensating it to 1-degree accuracy that takes professional expertise or else great patience on your part. Be sure you install anti-surge protection on the power line into the system.

Calibrating the electronics in an integrated system is no different from the calibration of individual instruments. There's more to do, though, and the tuning takes exceptional care, because inconsistencies accumulate and are compounded when they're allowed to enter the computations.

The most sophisticated systems offer calibration for twin through-hull speed transducers, for upwash, leeway, and heel error in the wind angle transducer, and for compass variation and deviation. Twin speed transducers will probably not give the same readings initially, but by turning one in its

tube until its reading matches the other, you can then do your mark-to-mark speed and distance calibration as for a single through-hull unit. If the system allows manual switching between transducers, you can calibrate each unit separately.

Upwash is the error created by the headsails' effect on airflow over the wind vane at the masthead. The sails bend the air to give it an apparent reading farther aft than it really is. Masthead rigs suffer significantly more than do fractional rigs. Upwash calibration brings the wind forward to

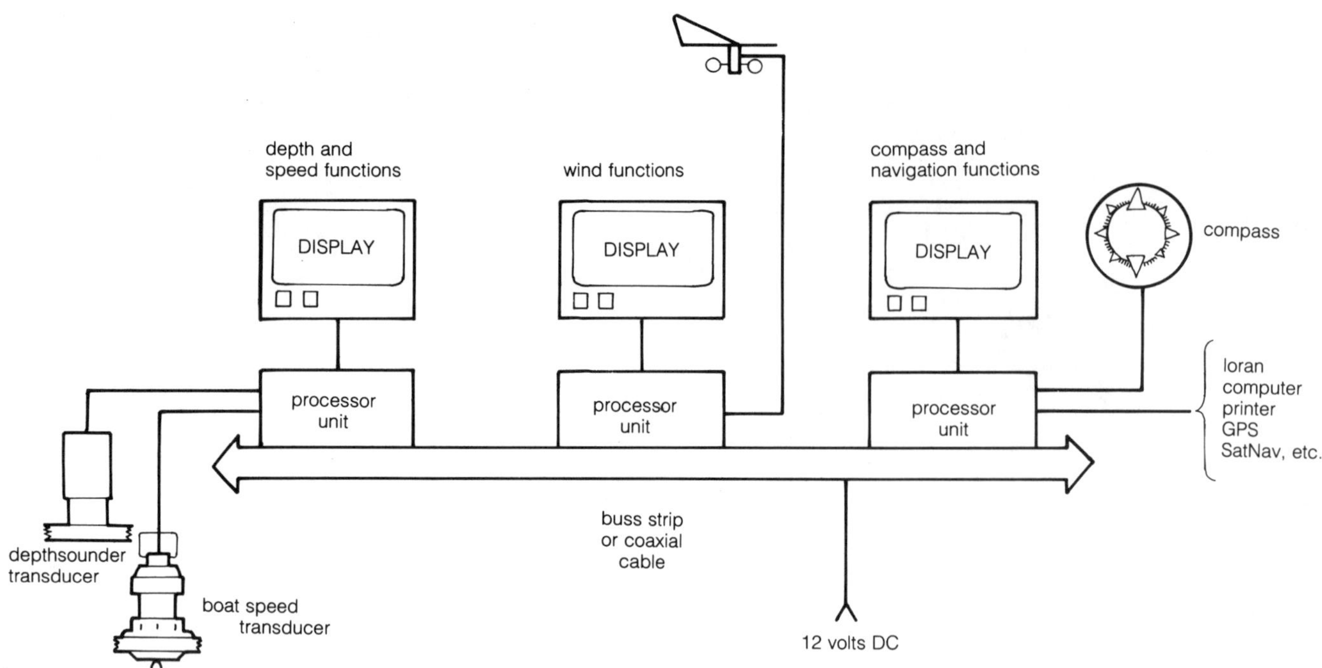

Network-type integrated systems transfer data along a common connector strip or cable, which is basically open ended. Limited only by the power of the chips within the processor units to communicate with one another, any number of processor units and displays can be added, each working independently as well as in conjunction with others.

where it would be if the genoa weren't hoisted. You can make this calibration on standard wind angle meters, too, simply by comparing the wind angles with and without the headsail hoisted, at various headings in a steady breeze.

Heel error is a predictable value, as is leeway for a given boat type, and these two factors are thus automatically dealt with through calibration. An electronic heel sensor—a potentiometer—is needed to complete the correction circuit.

Compass variation and deviation, automatically compensated for by Magnavox SatNav receivers, are not so simply adjusted in other instrument systems. In some, variation can be corrected at the factory if you tell the company where you will be sailing. Deviation can only be corrected on board, underway. It varies with heel, though, so simply motoring on different headings from a known point won't solve the problem to the degree needed on a sailboat with a full race electronics suite. Compensating magnets within the compass are too close, and can make the heel error worse on some points. Instrument manufacturers recommend larger magnets placed farther from the compass.

At a recent Ockam Instruments seminar, a potential buyer asked, if for his $7,000, his integrated system would come with installation and troubleshooting services included. The manufacturer answered him without blinking, "We have found that the best operating systems are the ones that are owner-installed." It was not a cop-out, but a statement of reality. Although these systems make sailing a boat with maximum efficiency easier, they also demand more effort on your part in installing, tuning, and understanding them.

Buying Tips

There is a wide range of choices in instrumentation, just as with any other piece of marine electronics.

A glimpse at the future navigator's station? Right now, this is what you see aboard a 60-foot super high-tech ocean racer for French singlehander Jean Yves Terlain. Top shelf, left to right: electrical panel; ICOM M-700 single sideband; Furuno 1700 digital radar; Shipmate RS5100 SatNav; engine gauges. Middle row: electrical panel; Magnavox MX 2400 integrated Satcom terminal, hooked to an antenna dish in a dome on the boat's stern; ICOM M5 VHF and printer for the Satcom; two Apple Macintosh displays for a navigation program called MacSea, which integrates performance and weather data to select a course for the next 24 hours; battery gauges. Bottom row: amp gauge; keyboard for the Satcom terminal; Robertson AP40 autopilot; Brookes & Gatehouse integrated instrument control units. Below the autopilot is a good old-fashioned barometer. The phone is left ashore when Terlain puts to sea. (Photo Roger Kennedy)

Prices vary, and so do accuracy, ruggedness, and dependability. At one time the cheap stuff was easy to spot: it had the plastic indicator housing and through-hull unit. Now almost every manufacturer's electronics come in plastic cases, only the material is now touted as lightweight rather than economical. Plastic has its advantages, though. It is a better choice for a watertight keypad than is aluminum or bronze with rubber seals, and for ease of maintenance. Bronze still makes a better through-hull tube, though.

Can you tell anything about an instrument by picking it up and looking at it? Sometimes. Does the transducer paddlewheel or spinner rotate freely? Blow on it. How much breath is needed to make it start to turn? Are the paddles well molded and balanced, with the magnets molded in and not just glued on? Is the transducer well potted in epoxy inside the through-hull tube? Are there connectors in the coax cable 5 to 10 feet from the through hull as well as at the indicator unit? Is the indicator itself well gasketed where it must face the elements? Are circuit boards inside coated with a waterproof sealer? Is matte safety glass used for the dial, rather than a plastic that could scratch? Are digital displays readable in bright sunlight and in the shade? Does the masthead wind speed and direction unit rotate at the faintest breath of air, or take 2 or 3 knots of wind to get cranking? How does it register on the indicator—instantly and wildly, or after a slight delay for damping and averaging? Are damping and averaging rates adjustable? Does the indicator give full 360-degree wind coverage?

You can compare spec sheets, too, of course. Power draw is usually expressed in microamps. Accuracy is most often seen as a percentage of true speed (such as ±2%) or in degrees of wind or compass angle (±1 degree). Sensitivity—how light a breeze or how low a boat speed is sufficient for reliable readings—depends on bearing quality and weight of moving parts. For masthead components, especially, this is important.

Dependability is best determined by looking at instruments aboard other boats in your harbor and asking the owners for a brief operational history. Be specific about failures, causes, repair time, warranty, and cost. Surveys of this type are more illuminating than all the dealer assurances you can get.

Where should you buy instruments and sailing computers? Is there any advantage in buying from a discounter? The question applies only to the simple, single-function equipment. This stuff is sold in large enough quantities to be offered at less than full markup. It is simple to install and test, and just about anyone can do the job. It thus makes sense to look for the lowest price on basic speedometers and wind instruments, assuming you're buying a well-known brand such as Datamarine, Standard, or Signet that's supported by the factory should something go wrong. For the more specialized gear, those electronics that require exceptional care in installation and calibration, it's worthwhile having a dealer around to help. Anyway, the integrated systems as a rule don't get discounted. Only two mail-order sources offer B&G, for instance. Pay the retail price, and expect retail service.

mail-order source	economy line	moderate line	premium line	integrated system
Coast Navigation	SR Instruments	Signet, Standard	Datamarine, Signet, VDO	Datamarine Link, Brookes & Gatehouse, Signet Smartpak, SR Nav 10
E & B Marine	EMS	Standard, Signet	Datamarine	
West Marine	Standard, Cybernet, Signet Landmark, SR Mariner	Signet	Signet, Datamarine	Signet Smartpak

Typical offerings of major mail-order electronics sources

Mail Order Discount Houses

The following list will serve as a starting point for the mail-order bargain hunter. The larger, better known firms are included, as are companies of regional interest. Many warehousers also operate showrooms for the benefit of local customers. Check with those in your area.

Some of the mail-order houses, such as Goldberg's Marine and E&B Marine, are general merchandisers of all manner of boat gear, while others specialize in marine electronics. The only concrete advantage in ordering from members of the latter category, such as Skipper Marine Electronics or American Marine Electronics, is that you can get more or better information and advice over the telephone from them. With that exception, wise shopping through the mail-order houses consists of finding the best price on the item you want at the time you want it. Keep in mind that shipping and handling charges vary from one company to another and can alter the bottom line, although the effect is not often dramatic on high-ticket marine electronic items.

Although, to our knowledge, the companies listed here all provide satisfactory service, the list is not intended as an endorsement, nor is any company's exclusion from the list intended as a condemnation. The addresses and telephone numbers were current as of May 1987.

American Marine Electronics
5725 Oleander Dr.
Wilington, NC 28403
(800) 243-0264

BOAT/US
880 S. Pickett St.
Alexandria, VA 22304
(800) 336-0226
BOAT/US offers discount prices to members only. Cost of membership is $17 per year and provides access to information, documentation, and insurance services.

Coast Navigation
1934 Lincoln Dr.
Annapolis, MD 21401
(800) 638-0420

Crook & Crook
2794 SW 27th Ave.
Miami, FL 33133
(800) 432-2381

Dayton Granger
P.O. Box 350550
Fort Lauderdale, FL 33335
(305) 463-3451

Defender Industries
P.O. Box 820, 255 Main St.
New Rochelle, NY 10801
(914) 632-3001

E&B Marine
P.O. Box 3138
201 Meadow Rd.
Edison, NJ 08818
(201) 819-4600

Goldberg's Marine
202 Market St.
Philadelphia, PA 19106
(800) 262-8464

International Marine Supply
819 Pompton Ave.
Cedar Grove, NJ 07009
(201) 857-2646

Manhattan Marine & Electric Co.
116 Chambers St.
New York, NY 10007
(212) 267-8756

Marine Center
1150 Fairview North
Seattle, WA 98109
(206) 682-1150

M.M.O.S.
3034 Malmo Dr.
Arlington Heights, IL 60005
(312) 952-8444
15219 Michigan Ave.
Dearborn, MI 48126
(313) 582-9480

National Marine Supply Corp.
Keystone Executive Plaza
12555 Biscayne Blvd., Suite 924
Miami, FL 33181
(800) 645-2565

Outer Banks Marine
Highway 70 East
P.O. Drawer 500
Beaufort, NC 28516
(800) 682-2225

The Ship's Store Distributors
501 E. Boston Post Rd.
Mamaroneck, NY 10543
(914) 698-0056

Skipper Marine Electronics
3170 Commercial Ave.
Northbrook, IL 60062
(800) 621-2378

Starboard Marine
3078 Lawson Blvd.
Oceanside, NY 11572
(800) 645-6541

Tom Taylor Co.
72 Fraser Ave.
Toronto, Ontario
CANADA M6K 3E1
(800) 268-1788

West Marine Products
P.O. Box 5189
2450 17th Ave.
Santa Cruz, CA 95063
(800) 538-0775

Manufacturers and Importers

The following table lists the manufacturers or North American importers of the marine electronic instruments and systems covered in this book.

	Depthsounders			VHF Radios	SSB Radios	Radio Direction Finders			Loran	SatNav	Autopilots	Radar	Anemometers	Wind Direction Indicators	Speedometers	Multiple Display Instruments	Integrated Systems
	Digital or meter	Flasher	Recording			RDF	ADF	VHF/ADF									
Alpha Marine Systems, 996 Hanson Ct., Milpitas, CA 95035. (408) 945-1155											✓						
Apelco, 1107 North Ward St., Tampa, FL 33607. (813) 877-9418	✓	✓	✓	✓		✓	✓	✓	✓	✓		✓					
Aqua Meter Instruments, 465 Eagle Rock Ave., Roseland, NJ 07068. (201) 228-3600	✓	✓	✓			✓									✓		
Atlantic Marine Div. IMI Industries, 40 Signal Rd., Stamford, CT 06902. (203) 357-8455	✓												✓	✓	✓	✓	✓
Bristol Electronics, 21 Cove St., New Bedford, MA 02744. (617) 997-3181	✓			✓										✓	✓		
Brookes & Gatehouse, 154 E. Boston Post Rd., Mamaroneck NY 10543. (914) 698-9330	✓					✓				✓			✓	✓	✓	✓	✓
Canadian Marconi, 2442 Trenton, Montreal, Quebec, CANADA H3P 1Y9. (514) 341-7630				✓	✓												
Celwave R.F., Route 79, Marlboro, NJ 07746. (201) 462-1880				✓	✓	✓											
Cetec Benmar, 3000 W. Warner Ave., Santa Ana, CA 92704. (714) 540-5120											✓						

| | Depthsounders | | | | | Radio Direction Finders | | | | | | | | | | |
Company	Digital or meter	Flasher	Recording	VHF Radios	SSB Radios	RDF	ADF	VHF/ADF	Loran	SatNav	Autopilots	Radar	Anemometers	Wind Direction Indicators	Speedometers	Multiple Display Instruments	Integrated Systems
Cinkel Industries, 6220 Shawson, Mississauga, Ont. CANADA L5T 1J8. (416) 676-1234											✓						
Coast Navigation, 1934 Lincoln Dr., Annapolis, MD 21401. (800) 638-0420			✓			✓											
Coastal Navigator, P.O. Box 55819, Seattle, WA 98155. (206) 363-9500	✓	✓	✓	✓		✓			✓				✓	✓	✓		
Combi Marine Div. IMI Industries, 40 Signal Rd., Stamford, CT 06902. (203) 357-8455	✓			✓	✓	✓	✓		✓	✓	✓			✓	✓		
Communications Associates, 200 McKay Rd., Huntington Station, NY 11746. (516) 271-0800					✓												
Compass Electronics, P.O. Box 366, Forest Grove, OR 97116. (503) 357-2111	✓	✓															
Coursemaster U.S.A., P.O. Box 486, Grand Central Station, New York, NY 10017. (212) 706-1395											✓						
CPT, 4336 NE 11th Ave., Ft. Lauderdale, FL 33334. (305) 564-1445											✓						
Cruising Designs, P.O. Box 151, Peabody, MA 01960. (617) 532-2712											✓						
Cybernet International, 100 Randolf Rd., Somerset, NJ 08873. (201) 560-0060				✓													
Danaplus, Inc., P.O. Box 321 LPO Niagara Falls, NY 14304. (716) 282-7706	✓												✓	✓	✓	✓	✓
Datamarine International, 53 Portside Dr., Pocasset, MA 02559. (617) 563-7151	✓		✓		✓				✓	✓			✓	✓	✓		

Company	Depthsounders			VHF Radios	SSB Radios	Radio Direction Finders			Loran	SatNav	Autopilots	Radar	Anemometers	Wind Direction Indicators	Speedometers	Multiple Display Instruments	Integrated Systems
	Digital or meter	Flasher	Recording			RDF	ADF	VHF/ADF									
D & D Electronics, 246 Main St., Albany, OR 97321. (503) 928-5375		✓	✓														
Digital Marine Electronics, 30 Sudbury Rd., Acton, MA 01720. (617) 897-6600									✓								
R.L. Drake, 540 Richard St., Miamisburg, OH 45342. (513) 866-3621				✓	✓												
Electro Marine Systems, 96 Fox Hunt La., E. Amherst, NY 14051. (716) 632-0100	✓	✓											✓	✓	✓	✓	
Energy Control, P.O. Box 127, Bellevue, WA 98009. (206) 455-4666											✓						
Epsco Marine, 411 Providence Hwy., Westwood, MA 02090. (617) 329-1500						✓						✓					
First Mate Marine Autopilots, 41 Kindred St., Stuart, FL 33494. (305) 286-4480											✓						
Furuno, 271 Harbor Way, S. San Francisco, CA 94080. (415) 873-9393		✓	✓	✓	✓	✓	✓	✓	✓	✓		✓				✓	✓
Jay Stuart Haft, P.O. Box 11210, Bradenton, FL 33507. (813) 746-7161	✓												✓	✓	✓		✓
Heath Co., Benton Harbor, MI 49022. (616) 982-3200	✓	✓	✓												✓		
Hull Electronics, 7563 Convoy Ct., San Diego, CA 92111. (619) 278-6140					✓												
ICOM America, 2112 116th Ave. NE., Bellevue, WA 98004. (206) 454-8155				✓	✓												

Company	Depthsounders			VHF Radios	SSB Radios	Radio Direction Finders			Loran	SatNav	Autopilots	Radar	Anemometers	Wind Direction Indicators	Speedometers	Multiple Display Instruments	Integrated Systems
	Digital or meter	Flasher	Recording			RDF	ADF	VHF/ADF									
II Morrow, P.O. Box 13549, 2345 Turner Rd., SE, Salem, OR 97309. (503) 581-8101									✓		✓						
Impulse Mfg., 329 Railroad Ave., Pittsburg, CA 94565. (415) 439-2072	✓	✓	✓												✓		
ImtraCorp. (Seafarer), 151 Mystic Ave., Medford, MA (617) 391-5660	✓	✓				✓										✓	
Intech, 282 Brokaw Rd., Santa Clara, CA 95050. (408) 727-0500								✓		✓							
International Marine Instrument Div. IMI Industries, 40 Signal Rd., Stamford, CT 06902. (203) 357-8455	✓	✓							✓	✓	✓	✓	✓	✓	✓	✓	✓
International Radiotelephone, 1819 S. Central #4, Kent, WA 98031. (206) 852-3436					✓												
Kenyon, 3 Republic Rd., North Billerica, MA 01862. (617) 667-6318	✓												✓	✓	✓	✓	✓
King Marine Radio, 5320 140th Ave. North, Clearwater, FL 33520-2846. (813) 530-3411	✓	✓	✓	✓	✓	✓	✓	✓	✓	✓					✓		
Lowrance Electronics, 12000 E. Skelly Dr., Tulsa, OK 74128. (918) 437-6881	✓	✓	✓														
Magnavox Marine Systems, 2829 Maricopa St., Torrance, CA 90503. (213) 618-1200									✓	✓							
Mars Electronics, Marine Sys. Div., 1301 Wilson Drive, West Chester, PA 19380. (215) 430-2500				✓					✓	✓	✓	✓					
Metal Marine Pilot, 2119 W. Mildred St., Tacoma, WA 98466. (206) 564-5902											✓						✓

| | Depthsounders | | | VHF Radios | SSB Radios | Radio Direction Finders | | | Loran | SatNav | Autopilots | Radar | Anemometers | Wind Direction Indicators | Speedometers | Multiple Display Instruments | Integrated Systems |
	Digital or meter	Flasher	Recording			RDF	ADF	VHF/ADF									
Micrologic, 20801 Dearborn St., Chatsworth, CA 91311. (818) 998-1216									✓								
Mieco Div. Polarad Electronics Corp., 70-20 Huntley Rd., Columbus, OH 43229 (614) 888-0557									✓								
Motorola Marine, 1301 E. Algonquin Rd., Schaumburg, IL 60196. (312) 397-1000					✓												✓
Nautical Electronics Co. (NELCO), 7095 Milford Ind. Rd., Baltimore, MD 21208. (301) 484-3284									✓								
Navidyne, 11824 Fishing Pt. Dr., Newport News, VA 23606. (804) 874-4488									✓	✓							
Navigation Instruments, 1934 Lincoln Dr., Annapolis, MD 21401. (301) 268-3120						✓											
Newmar, P.O. Box 1306, Newport Beach, CA 92663. (714) 751-0488						✓											
Ockam Instruments, 26 Higgins Dr., Milford, CT 06460. (203) 877-7453	✓												✓	✓	✓		✓
Paragon Electronics, P.O. Box 1456, Bellevue, WA 98009. (206) 454-0846	✓																
Plastimo USA, Inc., 6605 Selnick Dr., Rt. 100 Business Park, Baltimore, MD 21227. (301) 796-0002	✓										✓		✓	✓		✓	
Pilot Instruments, 177 Webster St., Suite A292, Monterey, CA 93940. (408) 649-1129																✓	
C. Plath, North American Div. of Litton Sys., 222 Severn Ave., Annapolis, MD 21403. 800-638-0428						✓	✓		✓	✓	✓						

	Depthsounders			VHF Radios	SSB Radios	Radio Direction Finders			Loran	SatNav	Autopilots	Radar	Anemometers	Wind Direction Indicators	Speedometers	Multiple Display Instruments	Integrated Systems
	Digital or meter	Flasher	Recording			RDF	ADF	VHF/ADF									
Racal Marine, 14778 NE 95th St., Redmond, WA 98052. (206) 882-0731					✓					✓	✓	✓					
Radar Devices, 2955 Merced St., San Leandro, CA 94577. (415) 483-1953												✓					
Radio Holland U.S.A., 6033 South Loop East, Houston, TX 77033. (713) 649-1048				✓	✓												
Ray Jefferson, Main & Cotton Sts., Philadelphia, PA 19127. (215) 487-2800	✓	✓	✓	✓		✓	✓		✓	✓	✓	✓	✓	✓	✓		
Raytheon Marine, 676 Island Pond Rd., Manchester, NH 03103. (603) 668-1600	✓	✓	✓	✓	✓	✓			✓	✓		✓					
RB II, 455 Cayuga Rd., Buffalo, NY 14225. (716) 632-7474	✓													✓	✓	✓	✓	
Regency Electronics, 7707 Records St., Indianapolis, IN 46226. (317) 545-4281	✓			✓		✓											
RF Communications Group, Harris Corp., 1680 University Ave., Rochester, NY 14610. (716) 244-5830				✓	✓												
Robertson-Shipmate, 400 Oser Ave., Hauppauge, NY 11788. (516) 273-3737											✓						
Rochester Instrument, 118 Airport Drive Suite 205, San Bernadino, CA 92408. (714) 824-0101																✓	
Ross Labs, 3138 Fairview Ave. E., Seattle, WA 98102. (206) 324-3950	✓	✓	✓														
SAIT, 33 Rector St., New York, NY 10006. (212) 422-6690				✓	✓												

| | Depthsounders | | | VHF Radios | SSB Radios | Radio Direction Finders | | | Loran | SatNav | Autopilots | Radar | Anemometers | Wind Direction Indicators | Speedometers | Multiple Display Instruments | Integrated Systems |
	Digital or meter	Flasher	Recording			RDF	ADF	VHF/ADF									
SBA Sideband Assoc., 2011 North Capital Ave., San Jose, CA 95132. (408) 942-1771					✓												
Scanmar Marine Prod., 298 Harbor Dr., Sausalito, CA 94965. (415) 332-3233											✓	✓					
Seatron Marine Electronics, 4312 Main St., Philadelphia, PA 19127. (215) 487-2800			✓		✓												
SGC, 13737 SE 26th St., Bellevue, WA 98005. (206) 746-6310					✓												
Sideband, 656 East Gish Rd., San Jose, CA 95112. (408) 275-1150					✓												
Signet Scientific, 3401 Aerojet Ave., El Monte, CA 91731. (213) 571-2770	✓												✓	✓	✓	✓	✓
Simrad, 2208 NW Market St., Seattle, WA 98107. (206) 789-6482	✓		✓			✓	✓		✓								
Si-Tex Marine Electronics, P.O. Box 6700, Clearwater, FL 33518. (813) 535-4681	✓	✓	✓	✓			✓		✓	✓		✓	✓	✓	✓		
SMR, Inc. 1401 NW 80th Ct., Miami, FL 33172. (305) 591-9433	✓	✓	✓	✓	✓												
SR Instruments, 600 Young St., Tonawanda, NY 14150. (716) 693-5977	✓												✓	✓	✓	✓	
SRD Labs, 381 McGlincey Ln., Campbell, CA 95008. (408) 371-2666								✓									
Standard Communications, P.O. Box 92151, Los Angeles, CA 90009. (213) 532-5300	✓			✓												✓	✓

	Depthsounders					Radio Direction Finders											
	Digital or meter	Flasher	Recording	VHF Radios	SSB Radios	RDF	ADF	VHF/ADF	Loran	SatNav	Autopilots	Radar	Anemometers	Wind Direction Indicators	Speedometers	Multiple Display Instruments	Integrated Systems
Stephens Engineering Assoc., 7030 220th St. SW, Mountlake Terrace, WA 98043. (206) 771-2182					✓				✓	✓							
Techsonic Industries, One Humminbird La., Eufaula, AL 36027. (205) 687-6615		✓	✓														
Telco Products, 25 Lumber Rd., Roslyn, NY 11576. (516) 484-6688				✓													
Telcor Instruments, 17785 Sky Park Circle, Irvine, CA 92714. (714) 250-1016	✓												✓	✓	✓		
Tele Comm Communications, P.O. Box 3232, Margate, NJ 08402. (609) 822-8588					✓												
Texas Instruments, P.O. Box 405, MS 3438, 2501 S. Hwy. 121, Lewisville, TX 75067. (214) 462-5220				✓	✓				✓	✓							
Tiller Master, 774 West 17th St., Costa Mesa, CA 92627. (714) 548-6111											✓						
Tracor Instruments Div., 6500 Tracor Lane, Austin, TX 78725. (512) 926-2800										✓							
Trade Consultants, Inc., P.O. Box 3422, Peabody, MA 01960. (617) 535-7360																✓	
Trans World Communications, 240 Pauma Pl., Escondido, CA 92025. (619) 747-1079					✓												
Trimble Navigation, 585 N. Mary Ave., Sunnyvale, CA 94109. (408) 730-2900								✓	✓								
Uniden Corp. of America, 6345 Castleway Ct., Indianapolis, IN 46250. (317) 842-0280		✓		✓													

	Depthsounders			VHF Radios	RDF	Radio Direction Finders			Loran	SatNav	Autopilots	Radar	Anemometers	Wind Direction Indicators	Speedometers	Multiple Display Instruments	Integrated Systems
	Digital or meter	Flasher	Recording			SSB Radios	ADF	VHF/ADF									
VDO Instruments, 980 Brooke Rd., Winchester, VA 22601. (703) 665-0100	✓												✓	✓	✓		
Viking Instruments, 400 Lake St., Kingston, MA 02364. (617) 426-5661												✓					
Wesmar, Box C3001, Bothell, WA 98041. (206) 481-2296			✓								✓	✓					
WH Autopilots, 655 NE Northlake Pl., Seattle, WA 98105. (206) 633-1830											✓						
Yaesu USA, 17210 Edwards Rd., Cerritos, CA 90701. (213) 404-2700				✓	✓												
Yazaki Corp., Route 4, Airport Rd., Princeton, MN 55371. (612) 389-2303	✓		✓										✓	✓	✓		

Comparison Charts for Depthsounders, VHF Radiotelephones, Cellular Phones, and Autopilots

Depthsounder Comparison Charts

Manufacturer	Model	Type	Max Depth (feet)	Transducer Frequency (kHz)	Power Out (watts RMS)	Features												Paper Size (inches)	Dimensions (inches) H × W × D	Weight (lbs.)	Voltage In d.c./a.c.	Suggested Retail Price
						Colors	Digital Readout Of Depth	Bottom Lock	Zoom	Memory	Water Temp.	Boat Speed	Anchor Watch	Inter-faces	Depth Alarm	Split Screen	White Line					
American Pioneer 611 2nd Ave. N.W. Seattle, Wash. 98107 206-789-7053	Side-scope A-310	CRT 10″	6,000	Dual	1,800	16	✓	✓	✓	✓	3 Transducers Adjustable				✓	✓	✓		9x11x13.5	22	12	$4,745
	Side-scope F-310	CRT 10″	6,000	Dual	1,800	16	✓	✓	✓	✓	3 Transducers Fixed				✓	✓	✓		9x11x13.5	22	12	$3,245
	Side-scope A-201	CRT 7″	6,000	Dual	1,800	16	✓	✓	✓	✓	3 Transducers Adjustable Waterproof				✓	✓	✓		9x9x13	22	12	$4,595
	Side-scope F-201	CRT 7″	6,000	Dual	1,800	16	✓	✓	✓	✓	3 Transducers Fixed Waterproof				✓	✓	✓		9x9x13	22	12	$3,095
	Fish-scanner 310	CRT 10″	4,000	Dual	1,800	16	✓	✓	✓	✓	360° Scan Sonar				✓	✓	✓		9x11x13.5	22	12	$7,445
	Fish-scanner 201	CRT 7″	4,000	Dual	1,800	16	✓	✓	✓	✓	360° Scan Sonar Waterproof				✓	✓	✓		9x11x13.5	22	12	$7,295
	Fish-scope 310	CRT 10″	6,000	Dual	1,800	16	✓	✓	✓	✓	Vertical			✓	✓	✓	✓		9x11x13.5	22	12	$2,745
	Fish-scope 201	CRT 10″	6,000	Dual	1,800	16	✓	✓	✓	✓	Vertical Waterproof			✓	✓	✓	✓		9x9x13	22	12	$2,595
Apelco Marine Electronics 1107 North Ward St. Tampa, Fla. 33607 813-877-9418	XVC-8000	CRT 6″	3,600	200	150	8	✓		✓		✓	✓	✓		✓				8x5x11	8.9	12	$995
	XCD 600	LCD/Recorder	600	200	150		✓	✓	✓		✓	✓	✓		✓		✓	4″	6.66x9.33x3.33		12	$895
	XCD-480	LCD	480	200	75		✓		✓		✓	✓	✓		✓	Optional pH			8x5.5x2.25	1.5	12	$589
	XCD 240	LCD	240	200	60		✓								✓				5.66x5.5x2.25	1	12	$299
	Ranger 420	Recorder	900	200	200		✓	✓	✓								✓	5″	10.75x13x5.5	6.6	12	$599
	Ranger 1650	Recorder	2,340	50	200		✓	✓	✓								✓	5″	10.75x13x5.5	6.6	12	$629
	FXL 400	Flasher	600	200	40	3	✓				✓				✓				5.75x7.75x3	1.7	12	$339
	Ranger 90	Flasher	90	200	40										✓				6.25x7.12x3.81	1.5	12	$189
	Ranger 360A	Flasher	360	200	40										✓				6.25x7.12x3.81	1.5	12	$189
Aqua Meter 465 Eagle Rock Ave., Roseland, NJ 07068 201-228-3600	530	LCD	400	100	188		✓						✓		✓				3.5x6.6x4.2	1.5	12	$357.50
	550	LCD	500	200	188		✓						✓		✓				3.5x6.6x4.2	1.5	12	$447.00
	301	LCD or FLASHER	800	200	200		✓				✓	✓	✓	✓	✓				7.5x9.75x3.75	2.5	12	$424.50
	376A	Flasher LED	100	200	188	Shallow water unit									✓				5.75x5.8x4.5	1.5	12	$357.50
	375	Flasher LED	240	200	88										✓				5.75x5.8x4.5	1.5	12	$295.50
	362	Flasher LED	200	200	125														5.75x5.8x4.5	1.5	12	$215
	365A	Flasher LED	360	200	125										✓					1.5	12	$295
Brookes & Gatehouse, Inc., 154 East Boston Post Rd., Mamaroneck, N.Y. 10543 914-698-9330	HS 911	LCD	1,000	171	50	Powerboat Dept, Speed, Log System Single Station						✓			✓				9x5x2.5	2.4	9-30	$1,150
	HS 911 PP	LCD	1,000	171	50	Dual Station						✓			✓				9x5x2.5	2.4	9-30	$1,940
	HS 911TP	LCD	1,000	171	50	Dual Station					✓	✓			✓	Plus Rudder and Planning Angles				2.4	9-30	$3,250
Coastal Navigator 4312 Stone Way N., Seattle, Wash. 98103 206-633-1950	C/NAV-9000	LCD Graph	600	200	50		✓	✓	✓		✓		✓		✓				5.5x6.75x2.75		12	$499.00
	DDS-400	LCD	875	200	20		✓	Talking Feature					✓		✓				4x5.75x8		12	$749.00
	DDS-300	LCD	499	200	15		✓								✓				3x5.3x4		12	$379.00
	DDS-99	Numitron Tube	750	200	20		✓							✓	✓				4x5.75x8		12	$499.98
	DDS-200+	LCD	499	200	14		✓				Round				✓				5.5 dia. x3.6		12	$399.98
	DDS-MFI	LCD	999	200	30		✓	SCAN			✓	✓	✓	✓	✓				3x5.3x4		12	$349.98

Continued

Continued

Manufacturer	Model	Type	Max Depth (feet)	Transducer Frequency (kHz)	Power Out (watts RMS)	Colors	Digital Readout Of Depth	Bottom Lock	Zoom	Memory	Water Temp.	Boat Speed	Anchor Watch	Interfaces	Depth Alarm	Split Screen	White Line	Paper Size (inches)	Dimensions (inches) H × W × D	Weight (lbs.)	Voltage In d.c./a.c.	Suggested Retail Price
Coastal Navigator Continued	DS-210	Flasher LED	200	200	10														5.5 dia. x4.5		12	$199.98
	DS-606B	Flasher Neon	360	200	80								✓		✓				6.25x8.5x4.5		12	$349.98
	DS-1010B	Flasher Neon	600	200	80								✓		✓				6.25x8.5x4.5		12	$399.98
Cybernet International 100 Randolph Rd., Somerset, NJ 08873 201-560-0060	CDF-500	LCD	500	200	100		✓				✓	✓	✓	✓	✓		Log		3.2x6.5x4.5	1.8	12	$549
	LX201	LCD	199	205	60		✓								✓				3.75x5.6x4.33	1.5	12	$375
	Sandpiper DL-2480	LCD	399	205	75		✓				Depth Offset				✓				4x7.33x6.1	2	12	$495
Datamarine International 53 Postside Dr., Pocasset, Mass. 02559 617-563-7151	International Offshore 3000	LCD	999	120	150		✓				Depth Offset				✓				4x9x7	3	12	$795
	Corinthian S200 DL	LCD	199	205	60		✓												5 dia. x3.4	2	12	$500
	Link 5100/ 6100	LCD	399	205	120		✓	Part of Link NAV System				✓		✓	✓				4.5x0.8x7	3	12	$1,700 $1,500
	Link 8000	LCD	999	120	200		✓	Part of Link NAV System					✓	✓	✓				4.5x6.8x3.8	6	12	$1,200
	Z-7000	Recorder LCD	1,000	192	312.5		✓		✓		✓	✓	Log		✓				7x8.8x3.7		12	$976
	Z-6100	Recorder LCD	600	192	62.5		✓		✓						✓				7x8.8x3.7		12	$612
Eagle P.O. Box 669 Catoosa, Okla. 74015 1-800-331-2301	Z-5000	Recorder LCD	400	192	50		✓												7x8.8x3.7		12	$534
	Z-15	Recorder	1,000	192	200						Waterproof							4"	9x12.4x6.5		12	$1,095
	Z-155	Recorder	3,500	50	200						Waterproof							4"	9x12.4x6.5		12	$1,095
	DDS-1	LCD	999	192	37.5		✓				Waterproof								3.5x6.3x6		12	$452
	DDS-3	LCD	999	192	37.5		✓				Waterproof								3 dia. x 2.25		12	$479
	IN-DASH-1	Flasher LED	60	192	5.6						Waterproof								4 dia. x 5.25		12	$386
	SST-WP	Flasher LED	60	192	37.5						Waterproof								7x8x6		12	$243
	SSI-WP	Flasher LED	120	192	37.5						Waterproof								7x8x6		12	$213
	SSI-WP	Flasher LED	120	192	37.5						Portable. Waterproof								9x8x9.75		Batt.	$228
	CV 990	CRT 9"	4,000	200/50	1,000	8	✓	✓	✓	✓	Dual Frequency		✓		✓	✓			10.33x9.6x12.5	21	11-27	$1,995
	CV 950	CRT 9"	4,000	200 or 50	1,000	8	✓	✓	✓	✓			✓		✓				10.33x9.6x12.5	19.5	11.27	$1,495
Echotec, Inc. 18686 142nd Ave. N.E. Woodinville, Wash. 98072 206-487-2450	CV550	CRT 5"	2,000	200 or 50	600	8	✓	✓					✓		✓				7.33x6.25x12	9.5	12	$1,295
	Bass & Bay	CRT 5"	500	50	600	8	✓	✓			Shallow water design				✓				7.33x6.25x12	9.5	12	$1,325
	CV220	TV	1,200	200 or 50	100	8	✓	✓					✓		✓				Uses your personal color TV set.		12	$995
Electro Marine System 96 Fox Hunt Lane East Amherst, NY 14051 716-632-0100	CKLD	LCD	300	200			✓			✓	✓	✓	✓	✓	✓				5 dia. x3.5	5	12	$695
	LDC	LCD	200	200			✓								✓				5 dia. x3.5	5	12	$395
	DSP	Analog	120	200			Needle Readout								✓				5 dia. x3.5	5	12	$310
	FCV-121	CRT 11"	3,840	Dual	1,000	8	✓	✓	✓	✓			✓	✓	✓		✓		15.9x16.3x17.7	65.1	20-45	$6,600
	FCV-201	CRT 10"	960	88/200	1,000	8	✓	✓	✓								✓		12.5x11.4x14.8	30.9	10-42	$3,950
	FCV-501	CRT 8"	1,000	200 or 50	500	8	✓	✓	✓								✓		9.6x11.7x12.2	21	11-40	$2,650
Furuno U.S.A., Inc. 271 Harbor Way So. San Francisco, Calif. 94083 415-873-9393	FCV-661	CRT 6"	1,152	200 or 50	100	8	✓							✓			✓		7.7x9.5x14.2	8.1	11-40	$1,400
	FCV-601	CRT 6"	1,000	200 or 50	100	Mono	✓										✓		7.7x9.5x11.6	6.2	11-40	$900
	FE-880	Recorder	5,400	28,50,200	500		✓	✓			Commercial unit				✓	✓	✓	8"	17.1x16.8x7	42	110-220	$5,850
	FE-881	Recorder	6,000	28,50,88,200	1,000		✓	✓							✓	✓	✓	8"	17.1x16.8x6.9	38	11-40	$3,650
	FE-808	Recorder	3,600	28,50,200	500		✓	✓							✓	✓	✓	8"	16.4x14.4x5.4	26.4	11-40	$2,750
	FE-606	Recorder	1,800	200 or 50	300		✓	✓							✓	✓	✓	6"	10x12.2x5.5	9.4	11-40	$1,400
	FE-6200	Re-corder	960	200 or 50	50	50		✓	✓						✓	✓	✓	6"	8.5x11.6x3	5.1	12	$1,100
	FE-4200	Recorder	960	200 or 50	50		✓	✓							✓	✓	✓	4"	6.5x8.5x2.8	4	12	$700
	FE-4000	Recorder	960	200 or 50	50												✓	4"	6.5x8.5x2.8	4	12	$550

Continued

Continued

Manufacturer	Model	Type	Max Depth (feet)	Transducer Frequency (kHz)	Power Out (watts RMS)	Colors	Digital Readout Of Depth	Bottom Lock	Zoom	Memory	Water Temp.	Boat Speed	Anchor Watch	Interface	Depth Alarm	Split Screen	White Line	Paper Size (inches)	Dimensions (inches) H × W x D	Weight (lbs.)	Voltage In d.c./a.c.	Suggested Retail Price
Furuno (Continued)	FE-502MKII	Recorder	2,460	50	300	Single-fish Detection at 167 Fathoms											✓	6"	13.25x13x7.75	30.5	12-32	$1,800
	FH-106	CRT 10"	5,400	60/150	1,000	8	✓		✓					Sector Scanning Sonar		Commercial					12-32	$14,300
	CH-12	CRT 11"	4,800	60/88, 150	1,200	8	✓		✓					Sector Scanning Sonar		Commercial					24/32	$20,600
	C5-50MKII	CRT 14"	5,400	55		8	✓		✓		✓	✓	360° Instantaneously Sonar, Commercial								110	$82,000
Humminbird Techsonic Industries, #3 Humminbird Lane, Eufaula, AL 36027, 1-800-633-1468	LCR4ID	Recorder LCD	120	200		2	✓	✓	✓		✓	Fish Displayed In Red.			✓	Waterproof			5.75x8.25x2	1.5	12	NA
	LCR 3004	Recorder LCD	120	200			✓								✓	Waterproof			5.75x8.25x2	1.5	12	NA
	LCR 4004D	Recorder LCD	480	200			✓	✓	✓						✓	Waterproof			5.75x8.25x2	1.5	12	NA
	LCR 8000	Recorder LCD	240	200			✓	✓	✓	✓	✓		Waterproof		✓				6.75x11.1x3	3.75	12	NA
	LCR 4000	Recorder LCD	120	200			✓	✓	✓	✓			Waterproof						5.75x8.25x2	1.5	12	NA
	LCR 8000D	Recorder LCD	1,500	200			✓	✓	✓	✓	✓		Waterproof		✓				6.75x11.1x3	3.75	12	NA
	Portable	Recorder LCD	120	200									Waterproof		✓				12.5x11.5x4.5	9.75	12	
	Super Sixty	Flasher LED	200	200					16° Cone Angle										6.25x7.25x6	2.2	12	NA
	Super Thirty II	Flasher LED	200	200					32° Cone Angle										6.25x7.25x6	2.4	12	NA
	Birdtrap	Flasher LED	200	200					Portable										9.75x8.75x14.5	8.5	12	
Impulse Manufacturing, 329 Railroad Ave., Pittsburg, Calif. 94567, 415-439-2072	8800	CRT 5"	1,000	200	120	5	✓	✓	✓		✓	✓	✓		✓	✓			6x7.5x10.25	6	12	$699.99
	6800	CRT 6"	1,000	200	120	Mono	✓	✓	✓		✓	✓	✓		✓	✓			6x7.5x7.25	5	12	$549.99
	6115	CRT 6"	1,000	200	120	Mono			✓				✓		✓				6x7.5x7.25	5	12	$399.99
	1800	Recorder LCD	480	200	70		✓	✓	✓						✓				5x6x2.8	1.5	12	$399.00
	XT-300	LCD	400	160	70				✓	Log, Distance To Go		✓			✓				2.75x4x1		12	$699.00
International Marine Instruments, Signal Road, Stamford, Conn. 06902, 203-357-8455	Micro-Line TK-910	Digital	300	200	60		✓		Depth Trend Indicator AGC/Offset Calibration						✓				4.4x4.4x1.4	1	12	$399
	Stowe Nav-Sounder 250	Digital	500	200	120		✓		Depth Trend AGC/Offset						✓				4.3x4.3x4	1	12	$595
Interphase Technology, 1336 Brommer St., Santa Cruz, Calif. 95062, 408-462-1755	SPECTRA	CRT 6"	1,500	200 or 50	150	8	✓	✓	✓		✓	✓	✓		✓				5.5x7.5x10.5		12	$995
	2020	CRT 6"	1,500	200 or 50	150	Mono	✓	✓	✓		✓	✓	✓		✓				5.5x7.5x7.8	5	12	$695
	LC-200	CRT 6"	500	200	85	Mono			✓					✓					5.5x7.5x7.8	5	12	$595
King Marine, 5320 140th Ave. N, Clearwater, Fla. 33520-2846, 813-530-3411	1350 A/B	Recorder	3,000	200/50	200		✓	✓	✓				✓	✓	✓		✓	6	9.6x4.5x3.9	8	12	$999
	1060 A/B	Recorder	1,990	200/50	200		✓				✓	✓	✓	✓	✓	✓	✓	4	7.4x12.2x3.5	6	12	$795
	900	LCD	360	120	200		✓	Fish Alarm Fish Size Indicator							✓	✓	✓		7.4x10.5x2.7	2.3	12	$549
	4000	CRT 8"	2,400	200/50	500	8	✓	✓	✓				✓			✓	✓		10.5x10.4x15.1	22	12	$1,899
	4060	CRT 4"	1,800	120	200	8	✓	✓	✓				✓		✓	✓	✓		8x9.5x11.6	7.25	12	$1,099
Koden International, 77 Accord Park Dr., Norwell, Mass. 02061, 617-871-6223	CVS-8811	CRT 11"	2,400	28, 40, 50, 68, 75, 200	1,000	8	✓	✓		✓	✓	✓	✓		✓	✓			13.9x11.7x16.7	44.2	11-40	$3,195
	CVS-8808	CRT 8"	1,200	28, 40, 50, 68, 75, 200	1,000	8	✓	✓			✓			✓		✓			10.5x9.3x13.8	22	11-40	$2,495
	CVS-88	CRT 8"	1,000	50 or 200	600	8	✓	✓								✓			8.5x8.7x12.1	19.9	11-40	$1,850
	CVS-101	CRT 6"	1,280	120	100	8	✓		✓		✓	✓			✓	✓			7.2x8.3x13.8	8.2	12	$1,250
	MVS-1	CRT 6"	1,280	120	100	Mono	✓		✓		✓	✓			✓	✓			7.2x8.3x10.4	5.8	12	$850
	ESR-120 Sonar	CRT 11"	5,600	60, 88, 136	1,500	8	✓					✓	360° Scan in 12 30°-Sectors						16.9x15.4x14.8	50.6	24, 32, 115	$16,495
Leisure Lectronics, 1220 Luke St., Irving, Texas 75061, 214-986-1806	8200 Computer Graph	LCD	160	200	250						✓	✓	✓	✓	8200 pixels				7x9x2	3.5	12	$800
	N200-D	LCD	200	200	75		✓									✓			3⅜ dia.		12	$300
	N100/60 AGC	Flasher	100/60	200	75														3⅜ dia.		12	$250
	N60/240 TC	Flasher	240	200	75														3⅜ dia.		12	$300
	N40/100 TC	Flasher	100	200	75														3⅜ dia.		12	$300
Lowrance Electronics, 12000 E. Skelly Dr., Tulsa, Okla. 74128, 918-437-6881	X-5	Recorder LC Graph	6,500	192	375		✓				✓	✓	✓	✓	✓		✓		7x8.8x3.7	2.6	12	$930

Continued

Continued

Manufacturer	Model	Type	Max Depth (feet)	Transducer Frequency (kHz)	Power Out (watts RMS)	Colors	Digital Readout Of Depth	Bottom Lock	Zoom	Memory	Water Temp.	Boat Speed	Anchor Watch	Inter-faces	Depth Alarm	Split Screen	White Line	Paper Size (inches)	Dimensions (inches) H × W × D	Weight (lbs.)	Voltage In d.c./a.c.	Suggested Retail Price
Lowrance Continued	X-4	Recorder LC Graph	999	192	75		✔		✔				✔		✔				7x8.8x3.7	2.6	12	$580
	X-16	Recorder	8,000	Dual	200					✔				✔			✔	4″	9x12.3x6.5	7.9	12	$1,070
	3200	LCD	999	192	75		✔			Operates to 70 MPH					✔				3.5x6.3x5.5	1.5	12	$458
	3400	LCD	999	192	75		✔			Operates to 70 MPH					✔				3 dia. x2.25	1.5	12	$470
	2330-C	Flasher	300	192	37.5				4″ Target Separation. Operates to 70 MPH										6.5x8.75x5.5	2.5	12	$270
	2260-C	Flasher	120	192	37.5				4″ Target Separation. Operates to 70 MPH										6.5x8.75x5.5	2.2	12	$235
	1240-A	Flasher	100	192	19.75					Operates to 70 MPH									4 dia. x5.25	1.43	12	$432
	1210-A	Flasher	60	192	5.6					Operates to 70 MPH									4 dia. x5.25	1.1	12	$350
Ray Jefferson Main & Cotton Sts., Philadephia, Pa. 19127 215-487-2800	T-2000	CRT 6″	480	200		7	✔		✔		✔	✔	✔		✔				5.5x8.75x9.5	5.5	12	$1,199.95
	T-2000A	CRT 6″	960	50		7	✔		✔		✔	✔	✔		✔				5.5x8.75x9.5	5.5	12	$1,249.95
	T-1500	CRT 6″	480	200		Gray	✔				✔	✔	✔		✔				5.25x9x8	5.5	12	$849.95
	T-1500A	CRT 6″	960	50		Gray	✔				✔	✔	✔		✔				5.25x9x8	5.5	12	$899.95
	T-480	CRT 6″	480	200		Gray	✔		✔				✔		✔				5.25x9x8	5.5	12	$599.95
	LCD-500	LCD Recorder	480	200	70		✔		✔		✔	✔	✔		✔		✔		4.25x6.5x2.25	2	12	$549.95
	MX-2550	Recorder	400	200			✔	Select Depth at any level.									✔	4″	6.5x2.75x8.75	3.5	12	$799.95
	MX-2500	Recorder	400	200				Select Depth at any level.									✔	1″	6.5x2.75x8.75	3.5	12	$699.95
	580LCD	Flasher	200	200				Flasher Type LCD Read out					✔		✔				4.5x4.5x4	1	12	$349.95
	515	Flasher	120								LED Flasher				✔				4.6x4.5x4	2	12	$249.95
	514	Flasher	120								LED Flasher				✔				4.6x4.5x4	1.75	12	$219.95
	513	Flasher	360										LED Flasher						4.6x4.5x4	1.75	12	$189.95
	6020	Flasher	120						Portable LED Flasher						✔				10x8x9.5	4.6	Batt.	$299.95
	245	LCD	300				✔			Will Operate at Boat Speed to 40 MPH	✔				✔				2.5x4.5x5	1	12	$399.95
	225	LCD	300				✔			Will Operate at Boat Speed to 40 MPH					✔				2.5x4.5x5	1	12	$299.95
	215	LCD	300				✔				✔	✔			✔				2.5x4.5x5	1	12	$449.95
Raytheon Marine	V-900	CRT 8″	2,640	50/75 200	600	8	✔	✔	✔								✔		11.5x9.5x12.5	28	12-40	$1,695
Raytheon Marine 676 Island Pond Rd., Manchester, N.H. 98107 603-668-1600	JFV-100	CRT 10″	9,999	28/50 75/200	1000	8	✔	✔	✔					✔	✔	✔	✔		2.14x12.51x15.12	26.4	12-40	$2,795
	JFV-216	CRT 11″	9,999	28/50 75/200	1000	16	✔	✔	✔	✔ Tape				✔		✔	✔		15x16.25x17.5	60	24,32/15	$4,035
	DC-2002/502	Recorder	5,940	200/50	200	Gray	✔	✔	✔					✔			✔	5″	10.75x13x5.5	6.9	12	$449/489
	DC-200/50	Recorder	320/1,152	200/50	110	Gray	✔										✔	5″	11.2x12.9x5.9	6.9	12	$289
	DC-50S	Recorder	320	50	110	Gray		✔		Shallow Water, wide angle							✔	5″	11.2x12.9x5.9	6.9	12	$289
	D-600	LCD	600	120	800		✔				✔	✔	✔	✔	✔				7x5.13x1.8	1.5	12-40	$579
	D-250	LCD	250	200	100		✔	✔							✔				4.25x7.8x6.1	1.75	12	$249
	F-360	Flasher	360	200	50										✔				6x7x4.75	2.6	12	$249
	F-720D	Flasher	720	125							Commercial Unit			✔					10.5x9.5x4.5	5.5	12-32	$799
Ross Laboratories 3138 Fairview Ave. E., Seattle, Wash. 98102 206-324-3950	Fineline	Recorder	6,000	200, 100,50, 28	1,000				✔		Survey-grade recorder						✔	8″	23x15.75x6	30	12-32 115	$3,450
Signet P.O. Box 5770 3401 Aerojet Ave., El Monte, Calif. 91734 818-571-2770	MK273	LCD	999	160	65		✔				✔		✔		✔				3.25x7.5x7.9	1.5	12	$895
	MK272	LCD	200	200	60		✔								✔				6x6x3.5	1.5	12	$395
	P-40	LCD	200	200	60		✔				✔				✔				4.1x5.33x1.06	1.5	12	$495
	P-100	LCD	200	200	60		✔			Log, Timer	✔	✔	✔		✔		Keel Offset		4.1x5.33x1.06	1.5	12	$1,195
Simrad, Inc. 2208 N.W. Market St., Seattle, Wash. 98107 206-789-6482	Skipper CS-115	CRT 10″	1,920	200 or 50	800	8	✔	✔	✔					✔	✔	✔			10.6x11.4x12		11-40	$2,395
	Skipper CS-116	CRT 11″	3,840	38, 50 200	800	8	✔	✔	✔					✔	✔	✔			14.9x14.7x16.1		11-40	$3,595
	Skipper CS-119	CRT 11″	7,500	38, 50, 200	1,000	8	✔	✔	✔	✔				✔	✔	✔			16.1x14.7x6.1		24	$3,995
	Simrad ES-380	CRT 11″	2,400	38	2,000	8	✔	✔	✔	✔	Fish Size Feature			✔	✔	✔			15.7x14.5x15.7		110	$28,900
	Simrad ES-700	CRT 11″	1,200	70	1,000	8	✔	✔	✔	✔				✔	✔	✔			15.7x14.5x15.7		110	$23,800
	Simrad S-113	CRT 12″	1,500	180	1,000	8	✔			360° Scanning Sonar				✔		✔			14.3x18x17		24	$10,975

Continued

Continued

Manufacturer	Model	Type	Max Depth (feet)	Transducer Frequency (kHz)	Power Out (watts RMS)	Colors	Digital Readout Of Depth	Bottom Lock	Zoom	Memory	Water Temp.	Boat Speed	Anchor Watch	Inter-faces	Depth Alarm	Split Screen	White Line	Paper Size (inches)	Dimensions (inches) H × W × D	Weight (lbs.)	Voltage In d.c./a.c.	Suggested Retail Price
Si-Tex Marine Electronics P.O.Box 6700 Clearwater, Fla. 33518 813-535-4681	HE-711	CRT 14"	3,000	Dual	1,000	7	✓	✓	✓	Auto Bottom Track				✓ Tape		✓ Dual Freq.			12.5x14x14.3	33	11-40	$2,395
	HE-710	CRT 14"	3,000	200 or 50	1,000	7	✓	✓	✓	Auto Bottom Track				✓ Tape		✓			12.5x14x14.3	33	11-40	$2,855
	HE-705	CRT 10"	1,200	200 or 50	500	7	✓	✓	✓	Auto Bottom Track						✓			11.5x10.5x11.75	17.75	11-40	$1,795
	HE-702	CRT 6"	800	107	100	7	✓		✓		✓		Auto Bottom Track Transducer Cone Coverage Indicator						5.75x7.37x10.5	9	12	$1,095
	Micronar M-700	CRT 4"	600	107	100	7	✓		✓		✓		Auto Bottom Track Transducer Cone Coverage Indicator						4.25x6x10	4.75	12	$895
	Micronar M-410	Recorder LCD	190	200	20		✓			Available with optional battery pack								LCD 588 pixels	5.5x4.4x1.2	1.2	12	$219
	Micronar M-430	Recorder LCD	190	200	20		✓				✓				✓			LCD 588 pixels	5.5x4.4x1.2	1.2	12	$309
	Micronar M-450	Recorder LCD	320	200	50	4	✓		✓	✓	Auto Track Range Image Scroll		✓		✓			LCD 2400 pixels	5.75x6x6.5	3.5	12	$595
	FL7	Flasher LED	480	200	50										✓				5x6.5x3.5	2.5	12	$269
	FL5	Flasher LED	600	200	50	3									✓				5x6.5x3.5	2.5	12	$399
	Micronar FL-6	Flasher LED	120	200	30								Portable option						4.37x6x2.25	1.1	12	$165
	Micronar FL-8	Flasher LED	240	200	50	3							Portable Option		✓				4.37x6x2.25	1.1	12	$305
	Micronar M-18	Flasher LED	120	200	50								In-Dash Mount						3.5 dia. 2x3x3	1.1	12	$255
	DL-1	LCD	199	200	30		✓								✓				3.25x4.25x4.58	1	12	$339
	DT-2	LCD	300	200	50		✓					Time			✓				3.5 dia.	.70	12	$359
	HE-203 MKII	Recorder	300	107	50				✓									3"	6.25x7.75x2.75	2.5	12	$469
	HE-357 MKIII	Recorder	800	50	100				✓									4"	6.8x10x3	4.4	12	$699
	HE-32 MKII	Recorder	1,200	107	500		✓	✓	✓			Dual Cone 18°+36°						4"	8x12x4.5	11	12	$999
	HE-301	Recorder	1,200	200 or 50	500		✓	✓	✓									6"	9.3x12.5x3.5	7	·12	$999
	Micronar HE-203	Recorder	120	107	50				✓									3"	6.25x7.75x2.75	2.5	12	$369
	Micronar M-310	Recorder	300	107	100				✓									4"	6.75x10x3	4.4	12	$699
	Micronar M-810	Sonar LCD	160	400	100					90° Side Scanning, Fish Depth & Range								LCD 2880 pixels	6.5x9.3x1.7	3.2	12	$1,195
SMR, Inc., 1401 N.W. 80th Ct. Miami, Fla. 33172 305-591-9433	CR-6000	CRT 6"	1,280	120	400	6	✓	✓	✓		✓	✓	✓	✓	✓				NA		12	$999
	SD100X	LCD	600	200	32		✓								✓				5.5x7.4x3.9	2.64	12	$279
	Mark III	Flashes neon	300	200	40										✓				5.5x7.4x3.9	2.6	12	$189
	D-600	Flasher Neon	600	200	40								In-Dash Mount		✓				NA			$229
SR Instruments 600 Young St. Tonawanda, N.Y. 716-693-5977	DOM-1	LCD	300				✓								✓				4.7 dia. x2.8		12	$375
	DM-1C	Dial	150					Needle							✓				4.7 dia. x2.8		12	$299
	DM-4	Dial	60					Needle											4.7 dia. x2.8		12	$235
	DM-5	Dial	60					Needle							✓				4.7 dia. x2.8		12	$265
	DM-3	Dial	60					Needle					✓		✓				4.7 dia. x2.8		12	$340
	PD-1A	LCD	300				✓	Keel Offset							✓				3.3x5.7x2.5		12	$349
	PD-1	LCD	200					Keel Offset											3.3x5.7x2.5		12	$259
	NAV-10	LCD	300				✓	Multi-function Sailboat Nav system					✓			✓				5x8x2		12	$2,895
	Group 3	Dial	60					Needles, Depth, Speed, Wind Pac.					✓		✓				6.1x17.6x1.8		12	$875
	Group 1	LCD Dial	300					Complete Depth, Speed, Wind Pac.					✓		✓				6.1x17.6x1.8		12	$1,520
Standard Communications Box 92151, Los Angeles, Calif. 90009-2151 213-532-5300	DS-2	LCD	400	200	40		✓				Powerboat		✓		✓				2.25x6.5x4.4	1	12	$447
	DS-1	LCD	400	200	40		✓				Sailboat		✓		✓				5 dia. x 4	1	12	$459
Westmar, Inc. Box C-3001 Bothell, Wash. 9804-3001 206-481-2296	55340	CRT	1,500	180	1,000	14	✓		✓	✓			360° Scan	✓			✓		3x8.75x11.5	4.5	12	$2,950

VHF Radiotelephone Comparison Charts

Manufacturer	Model	Output Power-Watts	Channel No. Transmit	Channel No. Receive	Channel No. Weather	Channel Display	Rotary Dial or Key-Pad Entry	Synthesized	Scan	Priority Ch.	Instant 16	Dimensions HxWxD inches	Suggested Retail Price
Apelco Marine Electronics 1107 North Ward St. Tampa, Fla. 33607 813-877-8418	VXL 5000	25	54	82	9	LCD	K.P.	✓			✓	3x9.75x8.87	$339
	VXL 7000	25	54	82	9	LCD	K.P.	✓	✓	✓	✓	3x9.75x7.8	$439
	VXL 9000 (Modulation meter. Memory)	25	54	82	9	LCD	K.P.	✓	✓	✓	✓	3x9.75x7.8	$539
	Clipper Jr. Handheld	4	54	82	9	LCD	K.P.	✓	✓	✓	✓	7.6x2.75x1.87	$469
Cybernet International 100 Randolph Rd. Somerset, NJ 08873 201-560-0060	CTX 2040	25	54	64	9	LCD	R.D.	✓			✓	2.8x7.5x9.75	$329
	CTX 3010	25	54	89	9	LCD	K.P.	✓		✓	✓	2.8x6.5x7.25	$359
	CTX 2050	25	44	50	5	LCD	R.D.	✓		✓	✓	3x8.5x11	$399
	CTX 2055	25	54	64	9	LCD	R.D.	✓	✓	✓	✓	3x8.5x11	$449
	CTX 2060 (40 channel memory. Program scan.)	25	54	89	9	LCD	K.P.	✓	✓	✓	✓	3x8.5x11	$649
	CTX 2088 (Removable 40 channel memory. Program scan.)	25	54	89	9	LCD	K.P.	✓	✓	✓	✓	3.25x9.5x12.6	$895
	CTX 2090 (40 channel memory.)	3-5	54	89	9	LCD	K.P.	✓	✓	✓	✓	7.25x2.6x1.5	$595
Furuno U.S.A., Inc. 271 Harbor Way So. San Francisco, Calif. 94083 415-873-9393	FM-55 Handheld	5	ALL	ALL	4	LCD	K.P.	✓	✓	✓	✓	6.7x2.9x1.7	NA
	FM-252 (Up to 5 pvt channels.)	25	ALL	ALL	4	LED	K.P.	✓		✓	✓	3.6x11.1x13.3	$775
King Marine 5320 140th Ave. N Clearwater, Fla. 33520-2846 813-530-3411	7000 (Hailer.)	25	55	81	9	LCD	K.P.	✓	✓	✓	✓	1.8x8.25x6.5	$499
	7200 (Telephone handset. Intercom.)	25	55	81	9	LCD	K.P.	✓	✓	✓	✓	1.8x8.25x6.5	$549
	7350 Handheld	3	55	82	8		R.D.					7x2.5x1.6	$329
ICOM, Inc. 2380 116th Ave. N.E. Bellevue, Wash. 98009-9029 206-454-8155	M-80	25	ALL	ALL	10	LCD	R.D.	✓	✓	✓	✓	3x8.5x8.3	$889
	M-55	25	ALL	ALL	10	LCD	R.D.	✓	✓	✓	✓	1.75x5.25x6	$529
	M-5 Handheld	5.5	ALL	ALL	10	LCD	K.P.	✓	✓	✓	✓	8.25x3x1.6	$629
	M-2 Handheld	5	78	78	4		R.D.	✓			✓	8.25x2.5x1.6	$499
	M-12 Handheld	1	12	12	4		R.D.	Diode matrix			✓	6.1x2.5x1.75	$439
Intech 282 Brokaw Rd., Santa Clara, Calif. 95052 408-727-0500	Mariner 90 (Commercial Grade. 10 Pvt channels.)	25	55	104	10	LED	K.P.	✓	✓	✓	✓	3.25x10.75x12	$1,138
	COM 160 (20 pvt Channel. Full duplex, Modular Const, Military spec, second receiver for ch 13 ops.)	25	55	104	10	LED	K.P.	✓	✓	✓		8x18x15 10x5x5	$6,500
Internationl Marine Intruments Signal Road, Stamford, Conn. 06902 203-357-8455	PRT-50	25	50	61	4	LED	R.D.	✓	✓	✓	✓	2.75x8x10.25	$399
	RT-78	25	78	84	4	LED	R.D.	✓	✓	✓	✓		$549
Ray Jefferson Main & Cotton Sts., Philadelphia, Pa. 19127 215-487-2800	5000M	25	49	59	9	LCD	R.D.	✓			✓	2.5x7x8	$399.95
	7878M	25	54	76	9	LCD	K.P.	✓	✓	✓	✓	2.5x7x8	$499.95
	789 Handheld	3+	55	80	9		R.D.				✓	6.8x3x1.75	$399.95
Radio-Holland 6033 So Loop E. Houston, TX 77033 713-649-1048	RT-2047 Sailor (Full Duplex. Up to 60 pvt channels. Selections call.)	25	ALL	ALL	ALL	LCD	K.P.	✓	✓	✓		4.5x8.8x10.7	$1,093
	RT-144 Sailor (Up to 5 pvt channels.)	25	ALL	ALL	ALL		Dial	✓			✓	8.66x12.6x6.5	$844
	RT-146 Sailor (Up to 5 pvt channels.)	25	ALL	ALL	ALL	LED	K.P.	✓			✓	8.7x4.7x3.58 Display handset	$1,094
Raytheon Marine 676 Island Pond Rd., Manchester, N.H. 98107 603-668-1600	Ray 33	25	54	82	9	LCD	K.P.	✓			✓	5.6x9.9x2.5	$369
	Ray 77	25	54	82	9	LCD	K.P.	✓	✓	✓	✓	5.6x9.9x2.5	$469
	Ray 78	25	55	83	8	LCD	R.D.	✓	✓	✓	✓	3.1x9.4x10	$799
	Ray 53A	25	54	57	4	LCD	R.D. & K.P.	✓			✓	4.25x9.75x12.25	$359
Regency Communication 7707 Records St. Indianapolis, Ind. 46226 317-545-4281	NC-7200 (Automatic Direction Finding VHF including antenna.)	25	55	71	4	LED	K.P.	✓	✓	✓	✓	3.6x11.2x11.1	$1,669
	NC-7100 (Receiver only. ADF.)			71	4	LED	K.P.					3.6x11.2x11.1	$1,219
	MT-6500 (10 pvt channel.)	25	64	86	10	Vac. Flu	K.P.	✓	✓	✓	✓	2.6x6.5x10.7	$469
	MT-5500XL (Program scan)	25	54	60	4	LED	K.P.	✓	✓			2.6x6.5x10.7	$419
	MT-5300	25	54	60	10	LED	K.P.	✓	✓	✓	✓	2.6x6.5x10.7	$369
	MT-5100	25	54	60	10	LED	K.P.					2.6x6.5x10.7	$329
Robertson-Shipmate 400 Oser Dr. Hauppauge, N.Y. 11788 516-273-3737	Shipmate RS-800	25	55	ALL	8	LED	R.D.	✓		✓		2.2x6.3x7.8	$695
Si-Tex Marine Electronics P.O. Box 6700 Clearwater, Fla. 33518 813-535-4681	Compact 55	25	49	63	9	LCD	KP	✓	✓		✓	NA	$399
	850	25	55	63	9	LCD			✓	✓	✓	2.45x6.6x9.1	NA
	891	25	55	80	9	LED	R.D.	✓	✓		✓	2.5x7.8x10.4	$569
	H-100 Handheld		55	78	9		R.D.					7.25x2.75x1.75	$329
SMR, Inc. 1401 N.W. 80th Ct. Miami, Fla 33172 305-591-9433	SL-9100 Handheld	5			5	LCD	R.D.	✓		✓	✓		$489.95
	ST-8000	25				LCD	K.P.				✓		$439.95
	ST-8100	25				LCD	K.P.	✓	✓	✓			$499.95
Standard Communications Box 92151, Los Angeles, Calif. 90009-2151 213-532-5300	Horizon Explorer (10 pvt. channels.)	25	48	48	6	LED	K.P.	✓			✓	2.25x7x10.25	$379
	Horizon Voyager (15 pvt. channels.)	25	55	55	6	LED	K.P.	✓			✓	1.75x7.8x8.66	$519
	Horizon USA Special Edition	25	50	50	3	LED	R.D.	✓			✓	2.75x7x9.5	$449
	Horizon Maxi (10 pvt. channels.)	25	55	55	6	LED	R.D.	✓	✓		✓	2.75x7.25x10.3	$469
	Horizon International	25	55	55	6	LCD	K.P.	✓	✓	✓	✓	3.75x8.75x13	$749
	HX-2005 Handheld (10 pvt. channels)	5	55	55	6	LCD	K.P.	✓	✓		✓	7.1x2.1x1.5	$599
Uniden Corp. of America 6345 Castleway Ct., Indianapolis, Ind. 46250 317-842-0280	MC-724	25	ALL	80	10	LCD	K.P.	✓	✓	✓	✓	7.4x10x2.4	$599.95
	MC-690	25	50	60		LCD	K.P.	✓	✓	✓	✓	7.25x9.8x2.25	$459.95
	MC-790	25		80	10	LCD	K.P.	✓	✓	✓	✓	7.25x9.8x2.25	$549.95
	MC 610	25	54	64	10	LED	R.D.	✓				7.25x10x2.25	$349.95
	MC 480	25	35	40	5	LED	R.D.	✓			✓	7.25x9x2	$329.95
	MC 990 Handheld	3	ALL	80	10	LCD	K.P.	✓	✓	✓	✓	7x2.6x1.75	$579.95
Yaesu USA 17210 Edwards Rd., Cerritos, Calif. 90701 213-404-2700	1903 Handheld	3	55	82	4	LCD	K.P.	✓	✓	✓	✓	7x2x1.75	$499

(Courtesy *Motor Boating & Sailing*, member Hearst Corporation, copyright 1986)

Cellular Phone Comparison Chart

Manufacturer	Number of Models	Memory Channels	Display Didgets	Portable	Backlit Keypad	Call Received Indicator	Last Telephone # Redial	Call Timer	Suggested Retail Price Range
ARA Manufacturing Co. 606 Fountain Pkwy. Grand Prairie, Tex. 75050 214-647-4111	3	10 16 16	8 16 16	1/3	✔	2/3	✔	2/3	$1,295-$1,895
AT&T 100 Southgate Pkwy. Morristown, N.J. 07960 201-898-8338	5	10, 10, 10, 11, 16	10	1/5	✔	✔	✔	✔	$1,295-$2,995
Advanced Cellular Technology 26010 Eden Landing Rd., Suite 5 Hayward, Calif. 94545 415-782-8780	1	10	7	✔	✔	✔	✔	✔	$1,900
Alpine Electronics 19145 Gramercy Pl. Torrance, Calif. 90501 213-326-8000	1	40	10		✔		✔		$1,900
Audiovox Corp. 150 Marcus Blvd. Hauppauge, N.Y. 11788 516-231-7750	2	30	10		✔		✔	✔	$1,799-$1,995
Blaupunkt Robert Bosch Corp. 2800 S. 25th Ave. Broadview, Ill. 60153 312-865-5200	1	30	10		✔		✔	✔	$2,500
Fujitsu America 3055 Orchard Dr. San Jose, Calif. 95134 408-946-8777	1	100	10		✔	✔	✔	✔	$2,000
General Electric Mobile Comm. Lynchburg, Va. 24502 804-528-7000	4	30	8	✔	✔	2/4	✔	✔	$1,400-$1,900
Mitsubishi Int'l 879 Supreme Dr. Bensenville, Ill. 60106 312-860-4200	3	30	10	1/3	✔		✔	✔	$1,900
Motorola, Inc. 1301 E. Algonquin Rd. Schaumburg, Ill. 60196 312-397-1000	10	100	10-14	3/10	✔	✔	✔	✔	$1,500-$6,000
NEC America, Inc. Mobile Radio Div. 4910 W. Rosecrans Ave. Hawthorne, Calif., 90250 213-973-2071	2	16	16	1	✔	✔	✔	✔	$1,400-$1,900
NovAtel Carcom 1923 Bomar St. Ft. Worth, Tex. 76103 817-332-3027	4	30	16		✔	2/4	✔	✔	$1,100-$1,500
Oki Telecom One University Plaza Hackensack, N.J. 07601 201-646-0011	4	100	14	1	✔	2/4	✔	✔	$1,700-$2,000
Panasonic Telecomm. Div. One Panasonic Way Secaucus, N.J. 07094 201-348-7933	4	10 30	10		✔		✔		$2,000
Radio Shack 1600 One Tandy Ctr. Fort Worth, Tex. 76102 817-390-3557	1	30	10		✔		✔		$1,300
Walker Telecomm. Mobile Comm. Div. 200 Oser Ave. Hauppauge, N.Y. 11788 516-435-0490	3	30	14		✔		✔		$1,600-$3,000

(Courtesy *Motor Boating & Sailing,* member Hearst Corporation, copyright 1986)

Autopilot Comparison Chart

Manufacturer/Source	System Layout	Heading Sensor	Command Options	Rudder Feedback	Drive Unit	Control Unit Features	Model/Price
	w-cockpit wheel t-tiller u-underdeck	c-repeater coil f-fluxgate p-photocell o-other	w-windvane 0-0180 3-0183 f-Furuno i-other interface	p-potentiometer s-simulated o-other	h-hydraulic linear l-electric linear r-electric rotary	p-proportional rate d-deadband a-autotrim r-and/or other remote helm	
Alpha Marine Systems (AMS) 996 Hanson Court Milpitas, CA 95035 408-945-1155	t,u t,u	f f	w,0 w	p p	l l	p,a,r	4404/$3295 3000/$2295
Autohelm America 40 Signal Road Stamford, CT 06902 203-357-8455	u w t t t	f f f f f	w w w w w	p,s s s s s	h,l,r r l l l	p,r,a p,r p,r p,r	6000/$2600 3000/$849 2000/$849 1000/$575 800/$395
Cetec Benmar 3320 W. McArthur Santa Ana, CA 92704 714-540-5120	u u u u u u	p p p p p p	0,3,f 0 0 0	p,s p p p p p	h,r h,r h,r h,r h,r h,r	p,r p,r p,r p,r p,r d,r	CC2100/$3880* CC220/$2570* CK210/$2150* CS21R/$1050* CS21/$625* Model 17/$1995
CPT, Inc. (Autopilot II) 4336 NE 11th Ave. Ft. Lauderdale FL 33334 305-564-1445	w	p	w	s	r	d,r,a	Autopilot II/$1295
Energy Control Corp. (Encron) P.O. Box 127 Bellevue, WA 98009 206-455-4666	u u	p p	0 0	p p	h,r h,r	p**,a,r p**,a,r,	77A/$2195 77C/$1995–$2195
First Mate Marine 41 Kindred Street Stuart, FL 33497 305-286-4480	w w t	p p p	w,0 0 w,0		l r l	p/d,r p/d,r p/d,r	WP-12/$895 PB-12/$895 TS-12/$695
King Marine 5320 140th Ave., N. Clearwater, FL 33520 813-530-3411	w w	f f	0,i 0,i		h r	p p, loran interface has dodge control	AP2000-10/$949 AP2000/$895

Manufacturer/Source	System Layout	Heading Sensor	Command Options	Rudder Feedback	Drive Unit	Control Unit Features	Model/Price
Marinex (Cetrek) 254 Church Street Pembroke, MA 02359 617-826-7497	u u u	f,c f,c f,c	0 0,w 0	p p p	h,r h,r h,r	p,r,d,a p,r,d,a p,r,d	7000/$4645* 727/$2750* 721/$2450*
Metal Marine (Wood Freeman) 2119 W. Mildred St. Tacoma, WA 98466 206-564-5902	u u	c o	0,3,w	o	h,r r	d,r d,r	500/$2385* 420/$2915
Navico North America 9 Old Windsor Road Bloomfield, CT 06002 203-242-3216	w t t	o o o	w w w	s s s	r l l	p,r p,r p,r	4000/$679 2500/$495 1600/$349
Furuno U.S.A., Inc. P.O. Box 2343 South San Francisco CA 94083 415-873-9393	u	f	0,f	p		p	FAP-50/$4295
Plastimo USA 6605 Selnick Dr. Baltimore, MD 21227 301-796-0002	t	p		s	l	p/d	AT50/$425
Raytheon Marine (NECO) 46 River Road Hudson, NH 03051 603-881-5200	u u u				h		8401/$1995 8401H/$2195 728/$3495
Robertson-Shipmate 400 Oser Avenue Hauppauge, NY 11788 516-273-3737	u u u	f,c f,c f,c	0,3 0,3 0,3	p,s p,s p	h,l,r h,l,r h,l	p,r p,r p,r	AP9/$4999* AP200DL/$3195 AP1000DL/$2195
Si-Tex Marine Electronics (Sea-Tex) P.O. Box 6700 Clearwater, FL 33518 813-535-4681	u	f	0,i	p	h	p,r	S-725/$2995
Tiller Master 774 W. 17th St. Costa Mesa, CA 92627 714-548-6111	w t	p p		** **	r l	p/d,r p/d,r	Wheelmaster/$995 Tillermaster/$659

Manufacturer/Source	System Layout	Heading Sensor	Command Options	Rudder Feedback	Drive Unit	Control Unit Features	Model/Price
II Morrow, Inc. (Sharp) P.O. Box 13549 Salem, OR 97309 503-581-8101	u u	f f	0 0	p p	h,r h,r	p,r p,r	B45M/$2195 B45/$1795*
Viking Marine 400 Lake St. Kingston, MA 02364 617-426-5661	t	p			l	d,r	Tiller Pilot/$499
Wagner Marine P.O. Box 1268 Bothell, WA 98011 206-823-1372	u u u u	f f/c f/c f/c	w,0 0 0 0	p p p	h,r h h h	p,r p,r p,r p,r	Micropilot/$2490 S-50/$2995 SE/$1765* MK IV/$3995*
WESMAR P.O. Box C3001 Bothell, WA 98041 206-481-2296	u u u	f f f		p p p	h,l,r h,l,r h,l,r	p,r p,r p,r	AP1100/$4100* AP900B/$2000* AP800/$1500*
W.H. Autopilots 655 Northlake Pl. NE Seattle, WA 98105 206-633-1830	u u u t	f f f f	0,3 0,3 0,3	p p p	h,l h,l l l	p,r p,r p,r p,r	P20/$3495 AP2/$3495 P3S/$2595 Tiller/$395

*Price does not include the drive unit
**Proportional steering calibrated by owner